电气工程基础双语教材

Electrical Engineering Fundamentals

U0655593

电路与电子技术

Electric Circuits and Electronics

吴青华　主编

季天瑶　樊利民　贺小勇　李梦诗　编写

中国电力出版社

CHINA ELECTRIC POWER PRESS

内 容 提 要

本书为电气工程基础双语教材。全书内容分为电路与电子技术两部分,共 11 章,主要内容包括绪论、电路及其基本物理量与元件、电压与电流定律、经典电路分析方法、直流稳态与暂态分析、正弦电路稳态分析、三相电路、二极管、双极型三极管、场效应管、运算放大器、逻辑电路等。每章均配有相应的课堂讨论环节,给学生提供机会进行学术英语的思维、口语表达及演讲的训练,帮助学生对所学知识活学活用,将理论与生产生活实际结合起来。

在部分章节中加入了与当代工程技术相结合的应用实例分析,阐述电路及电子技术在实际应用中的原理及技术最前沿的发展方向,有助于学生了解工程应用中各类型电路与设备及其分析设计方法,达到学以致用。为配合教学与读者自学,本书配有电子课件及习题答案。

本书可作为开设"电路与电子技术"双语或全英语教学的电气类、电子信息类、自动化类等专业教材,也可作为其他工科专业的选修教材及供电路与电子技术及电类科技英语有兴趣的读者阅读与参考。

图书在版编目(CIP)数据

电路与电子技术=Electrical Engineering Fundamentals:Electric Circuits and Electronics:汉、英 / 吴青华主编. —北京:中国电力出版社,2017.2(2025.1 重印)
电气工程基础双语教材
ISBN 978-7-5123-8433-0

Ⅰ.①电… Ⅱ.①吴… Ⅲ.①电路理论-双语教学-高等学校-教材-汉、英②电子技术-双语教学-高等学校-教材-汉、英 Ⅳ.①TM13②TN01

中国版本图书馆 CIP 数据核字(2015)第 243184 号

中国电力出版社出版、发行
(北京市东城区北京站西街 19 号 100005 http://www.cepp.sgcc.com.cn)
固安县铭成印刷有限公司印刷
各地新华书店经售

*

2017 年 2 月第一版 2025 年 1 月北京第五次印刷
787 毫米×1092 毫米 16 开本 15.75 印张 384 千字
定价 46.00 元

前　言

由于现代工科源于西方科学技术的发展成果，高等院校工科教学一贯基于西方体系，科技术语多来自英文翻译，因此本科阶段对工科学生进行科技英语读写训练非常必要。培养学生英文表达能力的同时，让学生了解知识的本源，为将来在国际期刊发表科技论文打下基础。然而由于课时量的限制，仅通过"科技英语"课程无法完成此任务，因此面向本科生开设双语课程成为大势所趋，并且急需针对中国学生语言和逻辑思维背景的双语教材与之相配合。

本书以中英文对照的方式编写，方便读者阅读。全面介绍了电路与电子技术的基本概念和基本定律、电子电路的基本应用及其分析设计方法，使学生获得必要的电路与电子学基本概念和基本定律知识，掌握基本分析、设计方法，了解电路与电子技术发展的概况，建立坚实的电路分析基础，并具备一定的电路设计能力。本书内容分为电路与电子技术两部分，主要内容包括电路及其基本物理量与元件、电压与电流定律、经典电路分析方法、直流稳态与暂态分析、正弦电路稳态分析、三相电路；二极管、双极型三极管、场效应管、运算放大器、逻辑电路。本书突出概念及定义，将电路与电子技术领域内的经典理论和分析方法，必要时通过类比的方法阐述，以便于读者理解。每章均配有相应的讨论环节，给读者提供机会进行学术英语的思维、口语表达及演讲的训练，帮助读者对所学知识活学活用，将理论与生产、生活实际结合起来。在部分章节中加入"实例分析"，通过具体的案例阐述电路及电子技术在实际中的应用，有助于读者了解工程应用中各类型电路的分析设计方法，学以致用。在相应地方介绍科学家的简历和背景故事，寓教于乐。

作为工科基础课之一，本书面向电气信息类、电子信息类、自动化类专业的本科生，适用于课程设置中设有"电路与电子技术"双语或全英必修课的专业，也适用于其他专业的"电路与电子技术"选修课，也可作为广大对电路与电子技术这门学科、对电类科技英语有兴趣的读者的科普读物。

全书共分为11章，第1、4、5、6、9章由季天瑶编写，第2、3章由樊利民编写，第7、8章由贺小勇编写，第10、11章由李梦诗编写。本书由华南理工大学吴青华教授主编并统稿，中科院深圳先进技术研究院英籍博士 Peter Z Wu 仔细阅读了本书英文部分并给出了专业的修改意见，香港大学侯云鹤副教授对本书提出了切实的审稿意见，在此表示衷心的感谢诚恳希望广大读者对本书的缺点和错误提出批评和指正。

注意：本书以英文为主，中文只用来辅助英文理解，为了排版美观，英文部分加入了部分拓展的课外知识，没有给出对应的中文翻译；将图表放入了中文部分，节约版面的同时，图表中给出了中文、英文对应图表名及图注，方便查阅。

编者

Preface

The advacements of modern industry and engineering stem from the gradual developments of science and technology in the western world. Teaching of engineering in colleges and universites has always been based on the western education system, and many terminologies are translated from English. Therefore, it is of great necessity to train engineering students in academic reading and writing using the English language, which will enable the students to gain a better understanding of the subject matter at its root level and later on prepare them for publishing research papers in international journals. However, due to the limited class hours, relying solely on the attendance of compulsory courses such as academic English is insufficient to fulfill such a task. A better approach is to open bilingual courses to undergraduate students. Hence, textbooks written in both English and Chinese, which fully consider the language and train of thought of Chinese students, are in urgent need.

To ensure the readability of the textbook for students, it is typeset so that Chinese content is displayed on the left and the corresponding English content is shown on the right. This book covers the basic concepts and principles of electrical circuits and electronics, practical applications of circuits, and the analysis as well as design of circuits. This book will enable readers to understand basic laws and principles, apply them to solve and design circuits, get to know the historical development of electrical circuits and electronics. It provides a sound base to enable readers to develop theoretical and experimental skills in the areas covered. The book consists of two parts—electrical circuits and electronics, including basic elements and circuits, nonlinear elements, voltage and current laws, circuit analysis methods, DC steady state and transient analysis, ac steady state analysis, three-phase circuits; diodes, BJT transistors, MOSFET transistors, operational amplifiers, logic circuits. The book focuses on basic concepts and definitions, presentation of theories and analysis methods in the areas mentioned above, where appropriate illustrations are provided to assist the readers' understanding of a concept. Discussions are also available in most chapters, which will help readers with the practical application of theories, as well as providing them with opportunities to practice their academic presentation skills. Some chapters also include examples explaining the application of electric circuits and electronics to solving real-world problems and demonstrate how the circuits are analyzed and designed in engineering. Relevant great scientists and engineers are introduced with their contributions highlighted.

As an elementary course for engineering students, this textbook is suitable for undergraduate students majoring in the subject of computer science, software engineering, automation, cybernetics, as well as other specialist subjects where the course of electrical citruits and electronics is mandatory and taught bilingual or in English. In addition, this textbook can also serve as an introduction for readers who are interested in electrical circuits and electronics and in academic English regarding electricity.

The book is organized into 11 chapters. The contributions made by each of the co-authors are listed as follows: chapters 1,4,5 and 9 were written by Dr. Tianyao Ji. Chapters 2 and 3 have been supplied by Limin Fan, while chapters 7 and 8 were written by Xiaoyong He. Chapters 10 and 11 were contributed by Dr. Mengshi Li. Prof. Qing-Hua Wu from South China University of Technology edited the completed book. Dr. Peter Z Wu from Shenzhen Institutes of Advanced Technology, Chinese Academy of Sciences has proofread the English version and made meticulous corrections to ensure its readability. Dr. Yunhe Hou from Hong Kong University has reviewed the book and provided useful comments. The authors would like to acconwledge the help and encouragement that have been provided to them during the writing of this book.

Readers are welcome to address comments and corrections.

the authors

目 录

Contents

绪 论
Introduction

古希腊人通过摩擦琥珀感受到了轻微电击，18 世纪初的欧洲人将产生静电视为神奇的魔术表演，18 世纪中期富兰克林设计的实验揭示了闪电本质上与人们发现的"电"一样，18 世纪末伏特首次发明了电池。从此，关于电的科学研究一直在飞速发展。电是一种大家最熟悉的能源，我们利用电能点亮城市，利用电能取暖度过漫长冬夜，跨过大陆长空传递信息，创造现代工业和引领数字革命。耳熟能详的领域如通信、计算机、控制系统、电磁学、电力系统等都与电有关。本书将介绍上述专业共同的基础——电路与电子技术，介绍电路和电子学的基本概念和基本定律、电路的基本应用及其分析设计方法。通过学习本书，读者可以了解电路与电子技术发展的概况，打下坚实的硬件基础，为在其他相关领域的进一步深造做好准备。

电气工程是工科的一个分支，几乎涵盖了所有与电有关的应用，包括电工学、电子学、电磁学。现今通常将电气工程分为如下领域：电子学、计算机、电力系统、通信、控制系统、信号处理、电磁学、微电子等。电子学主要研究电子的特性和行为，应用电子学的原理设计制造电路和电子器件。计算机是最常见的电子设备之一，其功能的实现离不开电路。电力系统主要涉及电能的生产、传输、配送和使用。通信是研究信息以电的形式进行传播的科学，比如电报、电话、无线广播、卫星电视、因特网等。控制系统通过传感器获得需要的信息，再根据这些信息利用电能控制其他设备动作。信号处理指对包含了信息的电信号进行处理，其目的是从中提取有用的信息。电磁学是一门研究电和磁相互作用的学科，主要研究电磁波、电磁场以及有关电荷、带电物体的动力学等。微电子研究和制造微小的电子电路和元件，电子电路集成在半导体材料上，使得电路尺寸减小，功耗降低，性能提升。

电路与电子技术是电气工程的基础课程。本书的前 6 章构成第一部分，主要介绍电路的基本概念和基础知识，读者将从最基本的原理开始，逐步学会复杂电路的分析和设计；后 5 章构成本书的第二部分，主要介绍各类基本非线性元件和简单电子电路，并应用在第一部分学到的电路分析方法对其进行分析。

The ancient Greeks experienced minor shocks of electricity by rubbing together amber; in the early 1700s, people from Europe regarded static electricity as an astonishing form of magic; in the mid-1700s, an experiment designed by Franklin revealed that lightning by its nature was the same as the electricity made by man; in the late 18th century, Volta invented the first battery in the world. From then on, research on electricity has been rocketing. We use electricity to light our cities, heat our flats during cold winter nights, communicate across vast lands and oceans, as well as create modern industries and bring about the digital revolution. Fields such as commutation, computers, control systems, electromagnetism and power systems can all trace their roots back to the discovery of electricity. This book will introduce the foundation for all these fields—electrical circuits, including the fundamental concepts and laws of electrical circuits and electronics, the applications of electrical circuits, together with how to analyze and design circuits. This book will help you to understand the construction of electrical circuits and electronics, gain solid background in hardware, and prepare you for further study in related fields.

Electrical engineering is a branch of engineering that involves the study and application of electricity, electronics, and electromagnetism, which has now been subdivided into a wide range of subfields, including electronics, digital computers, power systems, telecommunications, control systems, signal processing, electromagnetism, microelectronics, etc. Electronics studies the characteristics of electrons and designs electronic circuits and devices. Computers are one of the most commonly used electrical devices, where circuits are designed to allow computers to perform complex computations. Power systems deal with the generation, transmission, distribution and utilization of electric power. Telecommunication is the transmission of information in electrical form, such as telegraph, telephone, radio, satellite television, the Internet, etc. Control systems gather information via sensors and use electrical energy to control a physical process. Signal processing involves the processing of electrical signals that carry information, and its objective is to extract useful information from these signals. Electromagnetism is concerned with the study of interaction between electricity and magnetism, including electromagnetic waves, electromagnetic fields, physical interaction that occurs between electrically charged particles. Microelectronics relates to the study and manufacture of very small electronic designs and components, and circuits are integrated in semiconductor material, which ensures the circuits to have smaller size, lower power consumption and better performance.

Electrical circuit and electronics is a fundamental course for electrical engineers. The first six chapters of this book focus on the basic concepts and fundamental knowledge of electrical circuits. We will start from the fundamental principles and then learn to analyze more complex circuits. In the second part, from Chapters 7 to 11, we will learn about nonlinear electronic devices and solve electronic circuits, using the circuit analysis methods we have learnt in the first part.

第1章
Chapter 1

电路及其基本物理量与元件
Circuits and its Basic Physical Variables and Elements

1.1　电路

手电筒电路可以说是一个最简单的电路实例，它由一块电池、一只灯泡、一个开关和几条将这些元件连成一个闭合回路的导线组成，如图 1.1（a）所示。当开关闭合时，电池将化学能转换为电能，这些电能通过导线输送给灯泡，灯泡里的钨丝则将电能转换为光和热。当开关断开时，能量的转换和传输被切断，灯泡不发光。

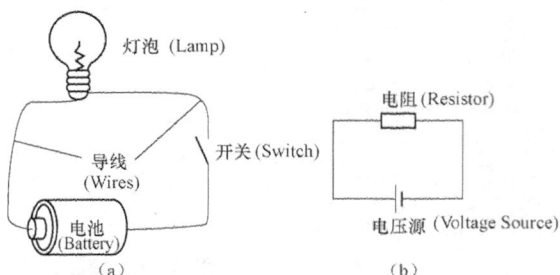

手电筒电路的电路原理图如图1.1（b）所示。在图1.1（b）中，电池用一个电压源表示，灯泡被抽象为一只电阻，开关被省略（因为通常我们只考虑电路闭合时的情况），导线用线段表示。电压源和电阻称为电路元件。除这两种外，常用电路元件还有电流源、电感和电容，它们的性质将在后续章节中介绍。电路分析的目的就是在已知电路构成的情况下，求解某些物理量（如流过电阻的电流或电容两端的电压等）或某些电路元件的参数（如电阻的阻值等）。

图 1.1　手电筒电路

（a）手电筒电路的示意图；（b）相应的电路图

Fig. 1.1　A torch circuit

(a) An illustration circuit; (b) The circuit diagram

1.2　基本物理量

1.2.1　电荷

电荷是电路中最基本的物理量。它是物质原子的基本属性，通常用 q 表示，单位为库仑（简称库，C）。一个电子所带的电荷为 -1.602×10^{-19} C。古希腊人所感受到的那种轻微电击，就是电荷作用的效果。在手电筒电路中，电压源迫使电荷流经导线和电阻。导线由一种导电性能极好的金属（例如铜）做成，这样电荷在通过导线时几乎不会受到阻碍。而电阻的导电性能则差很多，电荷流过电阻时会受到阻碍，将电能转换为热能。

1.2.2　电流

电流的定义为电荷流量随时间变化的速率。假设电荷和电流均是时间的函数并分别表示为 $q(t)$ 和 $i(t)$，则电流、电荷和时间之间的数学关系可表示为

$$i(t) = \frac{\mathrm{d}q(t)}{\mathrm{d}t} \tag{1.1}$$

电流的单位为安培（简称安，A），等同于库仑/秒。若已知电流，则在 t_0 时刻和 t 时刻之间移动的电荷可通过对式（1.1）两端进行积分求得，从而在 t 时刻的电量为

$$q(t) = \int_{t_0}^{t} i(t)\mathrm{d}t + q(t_0) \tag{1.2}$$

其中 t_0 时刻的电荷为已知。

电流是有方向的。在电路图中，我们通常用箭头表示电流的方向。大多数情况下，在进行电路分析时，并不清楚流过一段导线或流过一个元件的电流是什么方向。这就需要任意假设一个电流的参考方向。通常用不同的下标来区分流过不同元件的电流。如图 1.2 所示，标有 A、B、C、D 的小方框代表了电路元件，流过它们的电流分别用 i_1、i_2、i_3、…表示。

1.1 Circuits

A simple example of a circuit would be the circuit of an electric torch. It consists of a battery, a lamp, a switch, and wires connecting these elements into a closed loop, as shown in Fig. 1.1(a). The battery converts chemical energy into electrical energy; the electrical energy is then delivered to the lamp through the wires, and the tungsten filament in the lamp convert the electrical energy into heat and light. This happens when the switch is closed. When the switch is open, the flow of energy is disrupted and the lamp is turned off.

The circuit diagram of the electric torch is presented in Fig. 1.1(b). In this diagram, the battery is represented by a voltage source, the lamp corres ponds to a resistor, the switch is omitted as in most cases we care about a closed circuit, and the wires are represented by lines. Voltage sources and resistors are called circuit elements; other circuit elements include current sources, inductors, and capacitors, which will be discussed later in this book. The purpose of circuit analysis is to determine for a certain variable (such as the current flowing through a resistor or the voltage across a capacitor) or an element's parameter (such as the resistance of a resistor), when the configuration of a circuit is available.

1.2 Basic Physical Variables

1.2.1 Charge

The concept of electric charge is the most basic quantity in an electrical circuit. Charge is a fundamental property of atomic particles for which matter is made of; it is denoted by q and measured in the unit of coulomb (C). The charge on an electron is -1.602×10^{-19} C. The small shock felt by the ancient Greeks was the effect of electric charge. In the electric torch circuit, electrical charge is forced by the voltage source to flow through the wires and the resistor. The wires are made of an excellent electrical conductor, which is usually copper, so that the charge flows easily through them. The resistor, on the other hand, is not such an excellent conductor. The charge flowing through a resisor is opposed, causing the electrical energy to on be transformed into heat.

1.2.2 Current

Electrical current is defined as the rate of flow of electrical charge per unit time. Suppose the charge is expressed as a function of time denoted as $q(t)$ and the current is denoted by $i(t)$. The mathematical relationship between current, charge, and time is

$$i(t) = \frac{dq(t)}{dt} \tag{1.1}$$

The unit for currents is ampere (A), which is equivalent to coulomb per second (C/s). To find the charge transferred between time t_0 and t, we can integrate both sides of (1.1), and the charge at time t is

$$q(t) = \int_{t_0}^{t} i(t)dt + q(t_0) \tag{1.2}$$

where the charge at time t_0 is known.

Current flows along a direction. In a circuit diagram an arrow is often used to indicate the direction of a current. In many cases when we start to analyze a circuit, we may not know the actual direction of the current flowing through a certain conductor or in a particular circuit element. Therefore, we need to assume arbitrarily a reference direction for the current of interest. As a common practice, we use different subscripts to label the currents flowing through different circuit elements. For example in Fig. 1.2

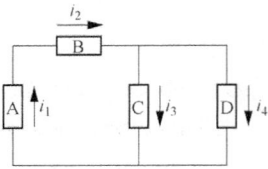

图 1.2　电流的参考方向

Fig. 1.2　A reference direction for each current

注意这些电流已被假设了参考方向。如果在求解该电路后我们得到电流 i_1 的值为负，比如 $-2\,A$，就说明电流的实际方向与我们假设的参考方向相反，实际上电流 i_1 是向反方向流的，大小为 2 A。

如果电流的大小和方向是恒定的，不随时间变化，则称为直流电流，缩写为 DC。图 1.3（a）所示为一大小为 2 A 的直流电流。若电流值随时间变化，且其方向也周期性反转，则称之为交流电流，缩写为 AC。交流电的波形多种多样，最常见的是正弦交流电流，如图 1.3（b）所示。其他波形还包括三角波和方波，如图 1.4 所示。本书中如不特别说明，交流电专指正弦交流电流。通常用大写字母 I 表示直流电流，用小写字母 i 表示交流电流。

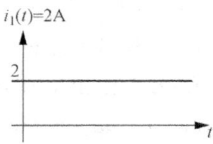
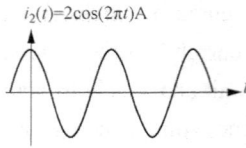
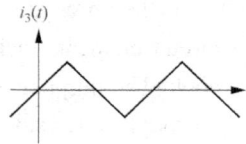
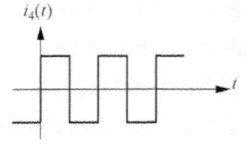

（a）　　　　　　　　（b）　　　　　　　　（a）　　　　　　　　（b）

图 1.3　两种常见电流

（a）直流电流；（b）正弦交流电流

Fig. 1.3　Two common types of current

(a) Direct current (DC); (b) Sinusoidally alternating current (AC)

图 1.4　其他波形的交流电

（a）三角波；（b）方波

Fig. 1.4　AC currents of various waveforms

(a) Triangular waveform; (b) Square waveform

1.2.3　电压

电荷在电路中的移动涉及能量的转移。比如在手电筒电路中，电池储存的化学能转移至灯泡，被灯泡吸收并转化为光能和热能。用于反映能量转移情况的物理量是电压，其定义为单位电荷流经电路元件所需的能量。其表达式为

$$v=\frac{\mathrm{d}w}{\mathrm{d}q} \tag{1.3}$$

式中，w 和 q 分别代表能量和电荷，v 代表电压，单位为伏特（简称伏，V），1 V=1 J/C。

如图 1.5（a）所示，a 点与 b 点处分别标注了正负号，则这两点之间的电压可表示为 v_{ab}。如果 a 点与 b 点处的极性符号交换，如图 1.5（b）所示，那么 b

（a）　　　　　　　　（b）

图 1.5　电压的表示与方向

（a）a 点为正极，b 点为负极；（b）a 点为负极，b 点为正极

Fig. 1.5　The representation and polarity of voltage

(a) The polarity at point a is positive, and the polarity at point b is negative;

(b) The polarity at point b is positive, and the polarity at point a is negative

点与 a 点之间的电压为 v_{ba}，a 点与 b 点之间的电压则为 $-v_{ab}$。显然，$v_{ab}=-v_{ba}$。在分析电路时，我们可能不知道某些电压的实际极性，这就需要先任意假设它们的参考极性。如图 1.6 所示，A～D 四个元件两端的电压分别由 v_1～v_4 表示，在进行分析计算之前先任意假设好它们各自的参考极性。如果经计算得到 $v_1=5\,V$，说明元件 A 两端电压值为 5 V，且电压极性与假设的参考极性相同。如果计算得到 $v_1=-5\,V$，则说明元件 A 两端电压的实际极性与参考极性相反。

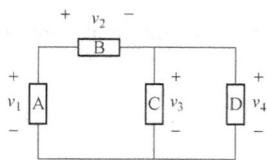

图 1.6　电压参考方向的设定

Fig. 1.6　Assigning reference polarities of voltages

the boxes labeled A, B, and so on represent circuit elements, and the currents flowing through them are denoted by i_1, i_2, i_3... Note that we have assumed a direction for each current. Suppose that after solving this circuit, we obtain a negative value for i_1, say−2 A. It means that the current actually flows in the direction opposite to the reference direction we have assumed. Thus, the actual current is 2 A flowing downward through element A.

If the direction of a current does not change with time, we call it a direct current, DC for short. An example of a DC current of 2 A is shown in Fig. 1.3(a). On the other hand, if a current varies with time and reverses direction periodically, it is called an alternating current, abbreviated as AC. An AC current may have various waveforms. Fig. 1.3(b) shows an AC current that varies sinusoidally with time. Other waveforms include triangular and square waveforms, as shown in Fig. 1.4.In this book, if not otherwise indicated, AC currents refer to sinusoidal currents. We normally use the upper case I to denote a DC current and a lower case i for an AC current.

1.2.3   Voltages

When charge moves through circuit elements, energy can be transferred. For example in the electrical torch circuit, the chemical energy stored within the battery is transferred to the lamp, which is then absorbed by the lamp and transformed to light and heat. We use voltage to describe energy transfer. Voltage is defined as the energy required to move a unit of charge through an element

$$v = \frac{dw}{dq} \tag{1.3}$$

where w and q represent energy and charge, respectively, and v represents voltage, which is measured in volts (V). One volt is equivalent to one joule per coulomb (J/C).

As shown in Fig. 1.5(a), points a and b are labeled by the plus and minus signs, respectively. Thus, the voltage between points a and b is expressed by v_{ab}. If the signs are switched so that point a corresponds to a negative polarity and point b to a positive polarity, as shown in Fig. 1.5(b), the voltage between points b and a is represented by v_{ba}, and the voltage between points a and b corresponds to $-v_{ab}$. Thus, $v_{ab} = -v_{ba}$. When analyzing a circuit, if the polarities of the voltages are unknown, we may start by assigning reference polarities for them, as we do for currents. As shown in Fig. 1.6, the voltages across elements A~D are labeled by v_1~v_4, and their reference polarities have been assigned before analysis. Suppose after calculation, we find that the voltage of element A is v_1=5 V; therefore, the voltage across element A has a magnitude of 5 V and its actual polarity is the same as what we have assigned. On the other hand, if we find that $v_1 = -5$ V, it implies that the voltage across element A has an opposite polarity to that shown in Fig. 1.6.

André-Marie Ampere (20 January 1775 – 10 June 1836) was a French physicist and mathematician who is generally regarded as one of the main founders of the science of classical electromagnetism. The unit of measurement of electric current, the ampere, is named after him. Ampère developed a mathematical and physical theory to understand the relationship between electricity and magnetism. He also discovered the principle that came to be called Ampère's law, which states that the mutual action of two lengths of current-carrying wire is proportional to their lengths and to the intensities of their currents.

和电流一样，电压也分直流和交流。直流电压不随时间变化，比如手电筒电池的输出电压可近似看作恒定保持在 1.5 V，在电路中就可以用直流电压源来表示。交流电压的值和极性随时间周期性变化，最常用的交流电压是正弦交流电压，比如家用插座的电压，可由方程 $v=220\sqrt{2}\cos(100\pi t)$ 表示。

1.2.4　功率与能量

1．功率

在本章第一节，手电筒电路中，能量从化学能转化为电能，再转化为光能和热能，这说明在电路分析中，能量也是非常重要的物理量。比如，日常生活中可能需要确定应该选一个多亮的灯泡，我们知道 15 W 的灯泡比 8 W 的亮，但是也更费电。因此计算功率和能量也是电路分析中必不可少的一部分。功率的定义为提供或吸收能量随时间变化的速率，单位为瓦特（简称瓦，W），该定义的数学表达式为

$$p=\frac{\mathrm{d}w}{\mathrm{d}t} \tag{1.4}$$

那么如何将功率和能量与电流和电压联系起来呢？回顾电流和电压的定义可知，电流是电荷随时间的变化率，电压是每单位电荷转移的能量。由此可知，电流和电压的乘积即为能量随时间的变化率，也就是功率。上述关系的数学表达式为

$$p=\frac{\mathrm{d}w}{\mathrm{d}t}=\frac{\mathrm{d}w}{\mathrm{d}q}\cdot\frac{\mathrm{d}q}{\mathrm{d}t}=vi \tag{1.5}$$

功率有提供功率和吸收功率之分。在根据式（1.5）来计算电路元件的功率之前，我们需

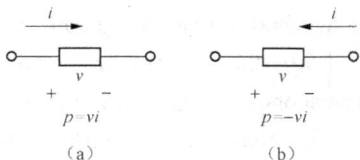

图 1.7　电功率的计算
（a）关联参考方向；（b）非关联参考方向
Fig. 1.7　Canculation of power
(a) Passive reference configuration;
(b) Against passive reference configuration

要确定该功率是由元件提供的还是被元件吸收的。如图 1.7（a）所示，箭头和正负号分别表示电流 i 和电压 v 的参考方向和极性。如果电流 i 从电压 v 的正极流入，则称电流和电压为关联参考方向，这种情况下通过公式

$$p=vi \tag{1.6}$$

计算功率。若结果为正，则元件吸收功率；反之，元件向电路中的其他元件提供功率。如果电流 i 从电压 v 的负极流入，如图 1.7（b）所示，则称电流和电压为非关联参考方向，通过公式

$$p=-vi \tag{1.7}$$

计算功率。同样，若结果为正，则元件吸收功率；反之，元件提供功率。

　　讨论：电流/电压参考方向/极性的意义

在开始电路分析之前，假设电流的参考方向和电压的参考极性，除了方便后续的求解（求解方法将在第 2 章中进行介绍），也帮助我们确定电路元件是在吸收还是提供功率。由于能量是守恒的，一个电路提供的功率与吸收的功率应时刻相等，这一原则可帮助我们检查电路分析的结果是否正确。

Similar to currents, voltages can also be regarded as DC or AC voltages. DC voltages are constant with time. The battery used in the electric torch can be considered as having a constant voltage of 1.5 V, thus it can be expressed by a DC voltage source in a circuit diagram. Ac voltages, on the other hand, change in magnitude and alternate in polarity with time periodically. The most widely used AC voltages are sinusoidal, such as the voltage of a household socket, which can be expressed as a function given by $v=220\sqrt{2}\cos(100\pi t)$.

1.2.4   Power and Energy

1. Power

In the first section of this chapter, we saw that in an electric torch circuit, the energy is transformed from chemical energy to electricity, then to heat and light, which implies that in circuit analysis, energy is also a variable of great importance. In some practical cases, we may need to know how 'bright' We need a lamp to be. We all know from experience that a 15 W lamp is brighter than an 8 W one, but it also consumes more energy. Thus, in circuit analysis, we also need to solve for power and energy. Power is the rate of energy transter, measured in watts (W). This relationship can be written as

$$p=\frac{dw}{dt} \tag{1.4}$$

To relate power and energy to current and voltage, let's recall the definitions of current and voltage. Current is the rate of flow of charge, and voltage is the energy transferred per unit of charge; therefore, the product of current and voltage is the time rate of energy transfer, i.e. the power. In mathematics, we can write this as

$$p=\frac{dw}{dt}=\frac{dw}{dq}\cdot\frac{dq}{dt}=vi \tag{1.5}$$

When we calculate the power of a circuit element using (1.5), we need to make sure whether the power is supplied or absorbed by the element. As shown in Fig. 1.7(a), the arrow and the $+/-$ signs indicate the reference direction of the current and the reference polarity of the voltage, respectively. If the current flows into the positive polarity of the voltage, such an arrangement is called passive reference configuration. In this case, the power is calculated by

$$p=vi \tag{1.6}$$

If a positive result is obtained, the element absorbs power; otherwise, it supplies power to other parts of the circuit. If the current enters the negative polarity of the voltage, as shown in Fig. 1.7(b), we say the current and the voltage are against passive reference configuration. Therefore, we compute the power in the following way

$$p=-vi \tag{1.7}$$

Likewise, a positive result implies the power is absorbed by the element, whereas a negative result shows that the power is supplied by the element.

Discussion: Why do we need to assign a reference direction/polarity?

Before we start to analyze a circuit, we need to assign a reference direction/polarity for a current/voltage of interest. By doing so, circuit analysis becomes easier (we will learn how to solve a circuit in Chapter 2). It also helps us to determine whether a circuit element is absorbing or supplying power. According to the law of energy conservation, the power absorbed in a circuit should equal to the power supplied at any given instant in time. This principle can be used to check whether our calculations are correct.

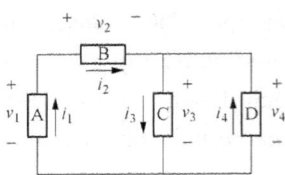

图 1.8　例 1.1 的电路图

Fig. 1.8　Circuit diagram for
Example 1.1

【例 1.1】　如图 1.8 所示，已知 v_1=6 V，$v_2=v_3=v_4=3$ V，$i_1=i_2=$ -2 A，经求解得到 $i_3=2$ A，$i_4=4$ A。请问这个求解结果正确吗？

解　利用能量守恒原则求解。首先根据已知条件分别计算出各元件的功率，再判断电路中吸收功率之和是否等于提供功率之和的绝对值（吸收功率为正，提供功率为负）。在计算各元件功率时，需注意其电流和电压是否为关联参考方向。各元件功率分别为

$$p_1 = -v_1 i_1 = -6 \times (-2) = 12 \text{ W}$$
$$p_2 = v_2 i_2 = 3 \times (-2) = -6 \text{ W}$$
$$p_3 = v_3 i_3 = 3 \times 2 = 6 \text{ W}$$
$$p_4 = -v_4 i_4 = -3 \times 4 = -12 \text{ W}$$

由计算结果可知，元件 A 和元件 C 吸收功率，其和为 18W，而元件 B 和元件 D 提供功率，其和为 -18W。显然，电路中各元件吸收功率之和等于提供功率之和的绝对值，这里的计算结果是正确的。

2. 能量

电路元件从 t_0 时刻至 t 时刻吸收或释放的能量可通过对功率的积分得到

$$W = \int_{t_0}^{t} p(t)\mathrm{d}t \tag{1.8}$$

能量的单位为焦耳（简称焦 J）。

【例 1.2】　求如图 1.7（a）所示的电路元件从 t_1=0 ～ t_2=∞ 转移的能量。已知其电流和电压的表达式分别为 $i(t)=2$ A，$v(t)=2\mathrm{e}^{-t}$ V。

解　首先计算功率。电流从电压正极流入，说明二者是关联参考方向，根据功率定义，有

$$p(t) = v(t)i(t) = 2\mathrm{e}^{-t} \times 2 = 4\mathrm{e}^{-t} \text{ W}$$

根据式（1.8）可求得能量为

$$W = \int_0^{\infty} 4\mathrm{e}^{-t}\mathrm{d}t = -4\mathrm{e}^{-t} \Big|_0^{\infty} = 4 \text{ J}$$

由于能量为正，可知该元件吸收能量。

补充阅读：数量级前缀

电路中物理量的数量级差异非常大。例如电力系统中，高压输电线的电压可达 500 千伏（kV），而在微电子中，电压通常只有几毫伏（mV）。因此，有必要将常用的物理量的数量级及其英文前缀对应起来，见表 1.1，以方便使用。

表 1.1　　　　　　　　　　　各常用前缀所代表的数量级

Table 1.1　　　　　　　　**Prefixes used for various orders of magnitude**

前缀	Prefix	缩写	数量级	前缀	Prefix	缩写	数量级
太	tera-	T	10^{12}	微	micro-	μ	10^{-6}
吉	giga-	G	10^{9}	纳	nano-	n	10^{-9}
兆	mega-	M	10^{6}	皮	pico-	p	10^{-12}
千	kilo-	k	10^{3}	飞	femto-	f	10^{-15}
毫	milli-	m	10^{-3}				

Example 1.1: For the circuit shown in Fig. 1.8, given that $v_1=6$ V, $v_2=v_3=v_4=3$ V, $i_1=i_2=-2$ A. Solving for i_3 and i_4 we obtain $i_3=2$ A and $i_4=4$ A. Is this answer correct?

Solution:

The problem is solved using the fact that all the circuits should obey the law of energy conservation. First we calculate the power for each element; then we find out the sum of power absorbed in the circuit and the sum of the power supplied in the circuit, respectively. If they have the same absolute value (power absorbed is positive whereas power supplied is negative), the answer is correct; otherwise, it is wrong. When calculating an element's power, we should pay attention to whether its current and voltage are of passive reference configuration. The power of each element is calculated by

$$p_1 = -v_1 i_1 = -6 \times (-2) = 12 \text{ W}$$
$$p_2 = v_2 i_2 = 3 \times (-2) = -6 \text{ W}$$
$$p_3 = v_3 i_3 = 3 \times 2 = 6 \text{ W}$$
$$p_4 = -v_4 i_4 = -3 \times 4 = -12 \text{ W}$$

The results show that elements A and C absorb power, the sum of which is 18 W; elements B and D supply a total power of -18 W. We can therefore safely conclude that the answer is correct.

2. Energy

The energy absorbed or supplied by a circuit element between time instants t_0 and t can be obtained by integrating the power

$$w = \int_{t_0}^{t} p(t)\mathrm{d}t \qquad (1.8)$$

Energy is measured in joules (J).

Example 1.2: For the circuit element shown in Fig. 1.7(a), find the energy transferred in the time interval from $t_1=0$ to $t_2=\infty$. The current and voltage are given by functions $i(t)=2$A and $v(t) = 2e^{-t}$ V, respectively.

Solution:

The first step is to calculate the power. As the current enters the positive polarity of the voltage, they are of the passive reference configuration, and the power can be calculated by

$$p(t)=v(t)i(t) = 2e^{-t} \cdot 2 = 4e^{-t} \text{ W}$$

According to (1.8), the energy transferred is given by

$$W = \int_{0}^{\infty} 4e^{-t}\mathrm{d}t = -4e^{-t}\Big|_{0}^{\infty} = 4 \text{ J}$$

As the energy is positive, the element absorbs the energy.

Supplementary reading: Prefixes

In circuit analysis, we encounter a wide range of values for currents, voltages, powers, and other quantities. In power systems, the voltage of high voltage transmission lines could reach 500 kilovolts (kV), while in microelectronics, the standard range of observed values is in the range of a few millivolts (mV). Therefore, for the purpose of convenience, the orders of magnitude and their corresponding prefixes are listed in Table 1.1.

1.3 基本电路元件

常用的电路元件主要有电压源、电流源、电阻、导线、电容、电感、二极管、晶体管、场效应晶体管等，本节将介绍前四种，其他元件将在后续的章节中逐一介绍。

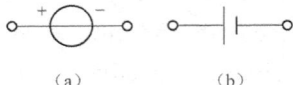

图 1.9 独立电压源的电路符号

（a）可用于表示直流或交流独立电压源；

（b）只可用于表示直流独立电压源

Fig.1. 9 Circuit symbols for independent voltage sources

(a) Used for DC or AC independent voltage sources;

(b) Used only for DC independent voltage soures

1.3.1 电压源

电压源，顾名思义是一种可以提供电压的元件。电压源有两种：独立电压源和受控电压源。独立电压源输出的电压，或者说两端之间的电压是恒定的，与电路中其他元件无关，与流过电压源的电流也无关。通常用如图 1.9（a）所示的电路符号表示独立电压源，正负号代表输出电压的实际极性，这个符号既可以表示直流电压源也可以表示交流电压源。有些情况下为强调某电压源是直流的，可用如图1.9（b）所示的符号来表示。

受控电压源又称为非独立电压源，其输出电压受电路中其他电压或电流的控制，或者说受控电压源两端的电压是电路中其他电压或电流的函数。受控电压源的电路符号是一个菱形，再用一对正负号表示输出电压的实际方向，如图 1.10 所示。图 1.10（a）中的受控电压源是电压控制电压源，它的输出电压是电路中另一个元件两端电压的函数。若 v_x=5 V，则电压控制电压源的输出电压为 15 V；若 v_x=10 V，则其输出电压为 30 V。图 1.10（b）所示为电流控制电压源，其输出电压是电路中某个电流的函数。

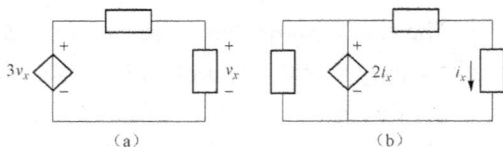

图 1.10 受控电压源

（a）电压控制电压源；（b）电流控制电压源

Fig. 1.10 Controlled voltage source

(a) Voltage controlled voltage source;

(b) Current controlled voltage source

1.3.2 电流源

同电压源一样，电流源也分为独立电流源和受控电流源两种。独立电流源为电路提供一个表达式恒定的电流，该电流的大小与电路中其他元件无关，与电流源两端的电压也无关。独立电流源的电路符号如图 1.11 所示其箭头表示电流的实际方向。该符号可用于表示直流或交流电流源。

图 1.11 独立电流源的电路符号

Fig. 1.11 Circuit symbols

for independent current sources

受控电流源也用菱形表示其中箭头表示电流的实际方向。受控电流源同样分为电压控制和电流控制两种，如图 1.12 所示，分别为电压控制电流源和电流控制电流源。

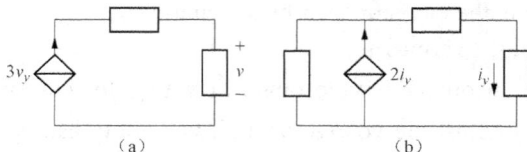

图 1.12 受控电流源

（a）电压控制电流源；（b）电流控制电流源

Fig. 1.12 Controlled current sources

(a) Voltage coutrolled curreut source; (b) Current controlled current source

1.3   Circuit Elements

Circuit elements include voltage sources, current sources, resistors, conductors, capacitors, inductors, diodes, BJT transistors, MOSFETs, etc. In this section, we will focus on the first four types of elements, and discuss the rest later on in the book.

1.3.1   Voltage sources

Voltage sources supply voltage to the rest of the circuit. We have two types of voltage sources: independent and controlled voltage sources. The output voltage of an independent voltage source, or the voltage between the two terminals of an independent voltage source, is constant, which is independent of the other elements in the circuit and the current flowing through it. The circuit symbol for an independent voltage source is given in Fig. 1.9(a). The pair of $+/-$ signs indicates the actual polarities. Sometimes we use the symbol shown in Fig. 1.9(b) to represent DC voltage sources.

Controlled voltage sources are also called dependent voltage sources; their output voltage is controlled by some other voltage or current in the circuit. In other words, the voltage across a controlled voltage source is a function of some other voltage or current in the circuit. The circuit symbol for a controlled voltage source is a diamond, with a pair of $+/-$ signs to indicate its actual polarities, as shown in Fig. 1.10. In Fig. 1.10(a) we have a voltage controlled voltage source, whose output voltage is expressed by $3v_x$, with v_x being the voltage across another element. If $v_x=5$ V, the voltage-controlled voltage source outputs a voltage of 15 V; if $v_x=10$ V, its output voltage is 30 V. Fig. 1.10(b) presents a current-controlled voltage source, the voltage across which is the function of a certain current in the circuit.

1.3.2   Current sources

Just like voltage sources, current sources are also categorized into independent and controlled current sources. An independent current source supplies a specified current to the circuit. The current of an independent current source is independent of the elements connected to it or the voltage across it. The circuit symbol for an independent current source is given in Fig. 1.11, which is also a circle but has an arrow to indicate the actual direction of the current. This symbol is used for both DC and AC independent current sources.

Controlled (dependent) current sources are represented by a diamond with an arrow indicating the actual direction of the current. The control variables can be either voltage or current. Fig. 1.12(a) shows a voltage controlled current source, and Fig. 1.12(b) presents a current controlled current source.

Alessandro Guiseppe Antonio Anastasio Volta (1745—1827) was an Italian physicist known for his invention of the battery in the 1800s. The invention of the battery plays an important role in Ohm's and Kirchhoff's further research work. He is also credited as the discoverer of methane. Volta's invention sparked a great amount of scientific excitement and led others to conduct similar experiments which eventually led to the development of the field of electrochemistry.

✓ **讨论：** 受控源的理解一共有四种受控源：

（1）电压控制电压源。

（2）电流控制电压源。

（3）电压控制电流源。

（4）电流控制电流源。

对于受控电压源，无论控制量是电压还是电流，其输出一定是电压。比如图 1.10（b）中的受控电压源，它的输出为 $2i_x$ V，而不是 $2i_x$ A。所以在判断一个受控源是电压源还是电流源时，要根据它的电路符号来判断，而不是根据控制量判断。

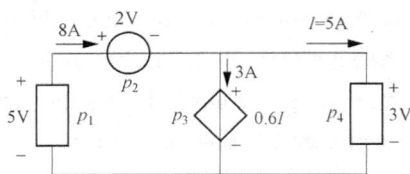

图 1.13　例 1.3 的电路图

Fig. 1.13　Circuit diagram for Example 1.3

【例 1.3】 求图 1.13 所示电路中各元件吸收或提供的功率。

解　$p_1 = -5 \times 8 = -40$ W，该元件提供功率；

$p_2 = 2 \times 8 = 16$ W，该元件吸收功率；

$p_3 = 0.6I \times 3 = 0.6 \times 5 \times 3 = 9$ W，该元件吸收功率；

$p_4 = 3 \times 5 = 15$ W，该元件吸收功率。

1.3.3　导线

导线用于将电路元件连接起来形成闭合回路。在电路图中，导线用实线表示。理想导线允许电荷畅通无阻地流过，无论流过理想导线的电流有多大，导线上任意两点间的电压均为零。实际中导线一般为铜导线，手电筒电路中的铜导线可以看成理想导线，然而在高压输电系统中，就不能再将铜导线视为理想导线了。在下一节介绍电阻的时候我们再具体解释原因。

若电路中有两点用理想导线连接起来，我们说这两点被短接，称为短路。在绝大多数情况下，短路将会引发事故，是需要避免的。如图 1.14 所示，电压源两端被短接，电压源两端电压为 v=5 V，而导线两端电压应为 v=0 V，显然这是矛盾的。在电路分析中，我们要避免这类矛盾。在实际中，如果将电池两端用一根导线短接，会有非常大的电流产生，电池储存的化学能将在非常短的时间内转化为导线中的热能，导线会烧融，电池会损坏。

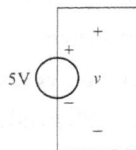

图 1.14　电源短路故障

Fig. 1.14　The short circuit of a voltage source

1.3.4　电阻

电阻也由导电材料制成，但其导电性能不如理想导线。电流流过电阻时会受到阻碍，电荷的移动速度降低。电阻通常是一个长圆柱形导体，如图 1.15（a）所示，电阻的阻值受圆柱体的截面积和长度影响，同时也与制成该电阻的导电材料有关。在圆柱形电阻的长度远大于其截面直径的情况下，

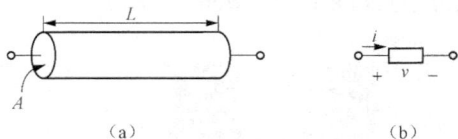

图 1.15　电阻

（a）圆柱形电阻示意图；（b）电阻的电路符号

Fig. 1.15　Resistors

(a) An illustration of a long cylinder resistor;

(b) Circuit symbol for resistors

Discussion: More about controlled sources

There are four kinds of controlled sources

- Voltage-controlled voltage sources.
- Current-controlled voltage sources.
- Voltage-controlled current sources.
- Current-controlled current sources.

The output of a controlled voltage source is voltage, no matter whether it is controlled by voltage or current. Take the controlled voltage source in Fig. 1.10(b) for example. Its output is $2i_x$ V, and not $2i_x$ A. Therefore, we should keep in mind to identify a controlled source by its circuit symbol, not by the control variable the source depends on.

Example 1.3: Compute the power absorbed or supplied by each element of the circuit in Fig. 1.13.

Solution:

$p_1 = -5 \times 8 = -40$ W, the element supplies power;

$p_2 = 2 \times 8 = 16$ W, the element absorb power;

$p_3 = 0.6I \times 3 = 0.6 \times 5 \times 3 = 9$ W, the element absorbs power;

$p_4 = 3 \times 5 = 15$ W, the element absorbs power.

1.3.3   Conductors

Circuit elements are connected by conductors to form a closed loop. In circuit diagrams, conductors are represented by solid lines. For an ideal conductor that allows the charge to flow freely through it, the voltage between any two points along the conductor is zero, regardless of the current flowing through it. We normally use copper wires as conductors. In the electric torch circuit, the copper wires can be considered as ideal conductors. However, in high voltage transmission lines, the copper wires can no longer be regarded as ideal conductors. We will came back to this when we discuss resistance.

When two points in a circuit are connected together by an ideal conductor, we say that the points are shorted together. Such a circuit is called a short circuit. In many practical applications, short circuits may cause accidents and should be avoided. For example in Fig. 1.14, the two terminals of a voltage source are shorted. According to the definition of the voltage source, its output voltage should be $v=5$ V. On the other hand, the definition of an ideal conductor requires that $v=0$ V. Apparently this is a conflict, which should be avoided in circuit analysis. In practice, if we connect the two terminals of a battery by a piece of copper wire, a very large current flows through the wire, the chemical energy stored in the battery is converted to heat in the wire at a very high rate, and the wire will probably melt and the battery will be destroyed.

1.3.4   Resistors

Resistors are also made from conductive materials, but their conductivity is not as good as ideal conductors. Current flowing through a resistor is hindered, which lowers the rate of the charge. Resistors often take the form of a long cylinder, as depicted in Fig. 1.15(a). The resistance of a resistor is influenced by its cross-sectional area and its length, as well as the material

电阻的阻值可通过下式计算

$$R = \frac{\rho L}{A} \tag{1.9}$$

式中，R 为电阻阻值，单位为欧姆（简称欧，Ω）；L 为圆柱形电阻的长度，单位为米（m）；A 为其截面积，单位为平方米（m²）；ρ 为导电材料的电阻率，单位为欧·米（$\Omega \cdot m$）。

电阻的电路符号如图 1.14（b）所示。任何导体都有一定的电阻，比如铜导线，在一般情况下其电阻可忽略不计，但是在高压输电系统中，由于输电线长达数千公里，所以铜导线的电阻不可忽略，不能再将其视为理想导线。

从导电性能角度看，材料可分为导体、半导体、绝缘体三种。导体的电阻率最低，电流容易通过。绝缘体的电阻率则非常高，电流几乎无法流过绝缘体。半导体的导电性能介于两者之间，在本书第二部分我们将看到半导体在电子技术中具有非常重要和广泛的应用。几种常见材料的电阻率见表 1.2。

表 1.2 　　　　　　　　　几种常见材料的电阻率

Table 1.2 　　　　　　　**Resistivity of common materials**

类型（Category）	材料（Material）	电阻率（Resistivity）
导体（Conductors）	银（Silver）	1.63×10^{-8}
	铜（Copper）	1.72×10^{-8}
	金（Gold）	2.44×10^{-8}
	铝（Aluminum）	2.73×10^{-8}
	钨（Tungsten）	5.44×10^{-8}
半导体（Semiconductors）	碳（Carbon）	3.5×10^{-5}
	硅（Silicon）	$10^5 \sim 1$，取决于掺杂浓度（10^{-5} to1, depending on impurity concentration）
	锗（Germanium）	47×10^{-2}
绝缘体（Insulators）	玻璃（Glass）	1×10^{12}
	特氟龙（Teflon）	1×10^{19}
	熔凝石英（Fused quartz）	$> 10^{21}$

used to construct the resistor. If the length of the resistor is much greater than the dimension of its cross section, the resistance is approximately given by

$$R = \frac{\rho L}{A} \tag{1.9}$$

where R is the resistance, measured by the unit of ohm (Ω), L is the length of the cylinder, A is the cross-sectional area, and ρ is the resistivity of the material used to construct the resistor, whose unit is ohm meter ($\Omega \cdot$ m). The circuit symbol for a resistor is given in Fig. 1.15(b).

Any conductor will have a certain resistivity. For example, in many cases the resistance of copper wires can be neglected. However, in a high voltage transmission system, as the transmission lines are of thousands of kilometers long, the resistance of copper wires should be taken into consideration; thus, we can no longer treat them as ideal conductors.

Materials can be classified as conductors, semiconductors, and insulators, according to their resistivity. Conductors have the lowest resistivity and currents flow easily through them. Insulators have very high resistivity and very little current may flow through them. Semiconductors fall between conductors and insulators. We will see in the second part of the book that semiconductors are very useful and have been widely applied in electronics. Table 1.2 lists the resistivity of some common materials.

▶ Problems

P1.1   The current flowing through a circuit element is given by

$$i(t) = 2e^{-t} \text{ A}$$

Find the net charge that passes through the element in the time interval from $t = 0$ to $t = \infty$.

P1.2   Suppose that a circuit element has $v(t) = 5e^{-t}$ V and $i(t) = 2$ A , and the reference   direction and polarity are shown in Fig. P1.1. Calculate the power for the circuit element and find   the energy transferred from $t=0$ to $t=\infty$. Is this element supplying or absorbing energy?

P1.3   Find the power absorbed or supplied by each of the elements in Fig. P1.2.

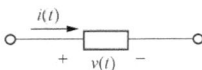

Fig. P1.1   Circuit diagram for P1.2

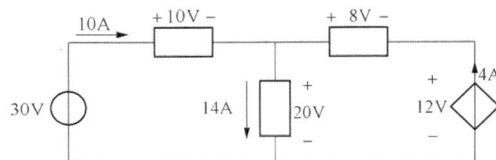

Fig. P1.2   Circuit diagram for P1.3

电压与电流定律
Voltage and Current Laws

上一章介绍了电路的基本概念和常用的电路元件。多个元件连接组成电路后，它们之间的电流、电压必然存在着制约关系。本章我们首先介绍反映各元件之间关系的基本电路定律，然后，结合欧姆定律，介绍电阻串并联电路的分析与化简。

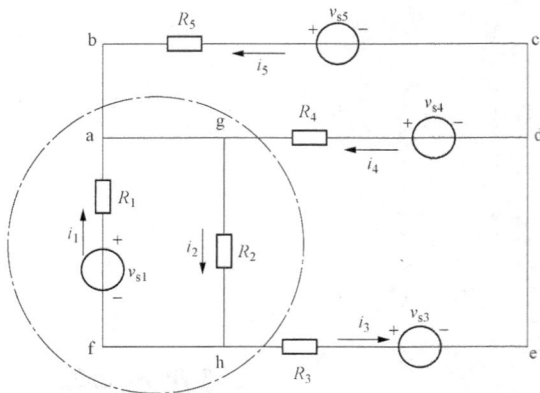

2.1 支路、节点和回路的概念

在学习电路基本定律之前，先介绍几个有关的术语。

支路：电路中的每一条分支都称为支路。一条支路流过一个电流，称为支路电流。图 2.1 中共有五条支路，bc、gd、af、gh、he，其中 bc、gd、af、he 中含有电源元件，称为有源支路；gh 支路不含电源元件，称为无源支路。

节点：电路中三条或三条以上支路的连接点称为节点。在图 2.1 中共有三个节点：a、d 和 h。

图 2.1　电路举例

Fig. 2.1　A circuit example

回路：电路中任一闭合路径称为回路。在图 2.1 中共有七个回路 a-b-c-d-g-a、a-g-h-f-a、g-d-e-h-g、b-c-e-f-b、a-d-e-h-f-a、b-c-d-e-h-g-a-b 和 b-c-d-g-h-f-a-b。

2.2 基尔霍夫电流与电压定律

2.2.1 基尔霍夫电流定律（KCL）

基尔霍夫电流定律（KCL）可表述为：在任一时刻，流入电路中任一节点的电流的总和等于从这个节点流出的电流的总和。在图 2.1 所示的电路中，对节点 a 或 g 可以写出：$i_1+i_4+i_5=i_2$。基尔霍夫电流定律给出了电路中连接到同一节点各支路电流之间的约束关系。

基尔霍夫电流定律所反映的是电流的连续性。电荷在电路中流动，不会消失也不会堆积（物质不灭定律），因此上式也可以改写成：$i_1+i_4+i_5-i_2=0$。

基尔霍夫电流定律也可以这样描述：在任一时刻，任一节点上，流入节点电流的代数和恒等于零。其数学表达式为

$$\sum i_k(t) = 0 \tag{2.1}$$

在上述求和式中，支路电流参考方向指向节点（流入节点）取"+"号，支路电流参考方向背向节点（流出节点）取"−"号。

基尔霍夫电流定律除了适用于电路中任一节点外，还可以应用于包围电路中某一部分的任一闭合面。在图 2.1 中，虚线闭合面包围了 aghf 围成的电路，有三条支路穿过该闭合面，支路电流分别是 i_3、i_4 和 i_5。根据物质不灭原理（电荷守恒），可以得出闭合面电流方程式为 $i_4+i_5=i_3$ 或 $i_4+i_5-i_3=0$。

所以基尔霍夫电流定律可以推广描述为：电路中，任意时刻流进任意一个封闭曲面的所有支路电流的代数总和等于零，其数学表达式为

$$\sum_{\text{流入封闭曲面支路电流}} i_k(t) = 0$$

In the previous chapter, we learnt some basic concepts of electrical circuits and some common circuit elements. Connecting circuit elements to form a circuit, the voltage and current of an element are related by certain laws. In this chapter, we will first introduce some basic laws used in circuit analysis. Afterwards, we will find out how to analyze and simplify a circuit based on Ohm's Law.

2.1 Concepts of Branches, Nodes and Loops

It is useful to clarify some terms before we set our hands on the circuit laws.

Branch: A branch refers to a part of a circuit in which a current flows through, which is therefore called the branch current. There are five branches in the circuit shown in Fig. 2.1, namely b-c, g-d, a-f, g-h and h-e. Branches b-c, g-d, a-f and h-e include sources, so they are called active branches; branch g-h contains no source and is thus called passive branch.

Node: In a circuit the point where three or more than three branches are linked together is called a node. In Fig. 2.1 we have three nodes which are a, d and h.

Loop: In a circuit any closed path that starting at a node, proceeding through circuit elements, and eventually returning to the starting node can be called a loop. We can find seven loops in Fig. 2.1, which are a-b-c-d-g-a, a-g-h-f-a, g-d-e-h-g, b-c-e-f-b, a-d-e-h-f-a, b-c-d-e-h-g-a-b and b-c-d-g-h-f-a-b.

2.2 Kirchhoff's Current and Voltage Laws

2.2.1 Kirchhoff's Current Law (KCL)

Kirchhoff's current law (KCL) can be expressed as: At any time instant, the sum of the currents flowing into a node equals to the sum of the currents flowing out of it. In the circuit shown in Fig. 2.1, for node a or node g, we have: $i_1+i_4+i_5=i_2$. KCL presents the relationship between the currents of all the branches connected to the same node.

KCL also shows that the current is continuous. As charges move through the circuit, they will not disappear or accumulates, according to the law of conservation of matter. Therefore, the above equation can be rewritten as: $i_1+i_4+i_5-i_2=0$.

KCL can also be expressed in this way: At any time instant at any node, the algebraic sum of the currents flowing into the node is always zero. Mathematically, it can be described as

$$\sum i_k(t) = 0 \tag{2.1}$$

In the above equation, if the reference direction of a branch current points to the node, indicating that the current flows into the node, we assign '+' for the current; otherwise, '−' is assigned for the branch current.

As KCL can be applied to any nodes in a circuit, it can also be applied to any loop of the circuit. In Fig. 2.1, the circle drawn in dashed line covers the loop of a-g-h-f. Three branches are connected to this loop and the branch currents are i_3, i_4 and i_5, respectively. According to the law of conservation of charge, for the loop we have $i_4+i_5=i_3$ or $i_4+i_5-i_3=0$.

Therefore, KCL has an extension version of definition: At any time instant in a circuit, the algebraic sum of the branch currents flowing into a loop equals zero, and it can be expressed as

$$\sum_{\text{Branch current flowing into the loop}} i_k(t) = 0$$

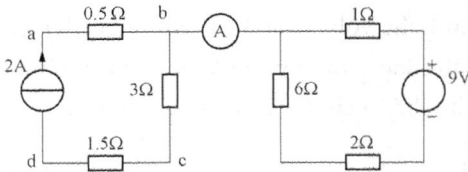

图 2.2　例 2.1 的电路图

Fig. 2.2　Circuit diagram for Example 2.1

【例 2.1】电路及元件参数如图 2.2 所示，求电流表 A 的读数。

解　该电路可以根据基尔霍夫电流推广定律求解。

图中节点 a、b、c、d 围成的是一个闭合曲面，且只有一条支路连接这个闭合面，因此电流表的读数应该为零。

2.2.2　基尔霍夫电压定律（KVL）

基尔霍夫电压定律：任何电路中，任一时刻沿任意一个回路绕行一周所有支路电压（降）的代数和恒等于零，其数学表达式为

$$\sum v_k(t) = 0 \tag{2.2}$$

应用 KVL 列方程时，首先需要指定一个回路绕行方向（顺时针或逆时针）。当支路电压的参考方向与回路绕行方向一致时（电压降），在求和式中取"+"号；当支路电压参考方向与回路绕行方向相反时（电压升），在求和式中取"−"号。

图 2.3 是某电路的一个回路，绕行方向为顺时针方向。按图中所指定的各元件电压的参考方向，根据式（2.2）可以写出

$$v_{ab} + v_{bc} + v_{cd} + v_{de} - v_{fe} - v_{af} = 0$$

或

$$v_{ab} + v_{bc} + v_{cd} + v_{de} = v_{fe} + v_{af}$$

基尔霍夫电压定律给出了电路中组成回路的各支路电压之间的约束关系，是能量守恒定律在电路中的体现。

基尔霍夫电压定律除了适用于闭合的电路外，还可以应用于不闭合的电路或某一段电路。例如，图 2.4 中有 6 个元件，a、d 之间的电压有两种计算方法

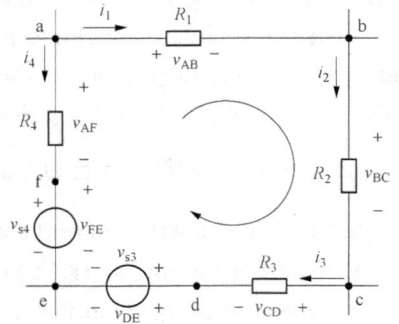

图 2.3　基尔霍夫电压定律电路举例

Fig. 2.3　An example of applying KVL

$$v_{ad} = v_{ab} + v_{bc} + v_{cd} \text{ 或 } v_{ad} = v_{af} + v_{fe} + v_{ed}$$

任何电路中，任意两节点之间的电压，与计算时所取的路径无关。

综上所述，基尔霍夫电流定律反映了电路的结构对节点上各支路电流所加的约束关系；而基尔霍夫电压定律反映了电路结构对回路中各支路电压所加的约束关系。必须指出的是，以上在对基尔霍夫定律的讨论中，对各支路元件并无要求，也就是说基尔霍夫定律只与电路的结构有关，而与元件的性质无关，故适用于任何线性或非线性电路。

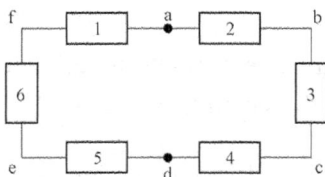

图 2.4　基尔霍夫电压定律的推广

Fig. 2.4　Extension of KVL

讨论： 电源的等效合并

1. 电压源的串联等效

用 4 节 1.5V 电池如何获得 6V 的电源电压？

Example 2.1: For the circuit shown in Fig. 2.2, find out the reading of the ammeter.

Solution: The circuit can be solved by applying the extended version of KCL. In this circuit, nodes a, b, c and d form a closed loop and there is only one branch connected to this loop. Therefore, the reading of the ammeter is zero.

2.2.2   Kirchhoff's Voltage Law (KVL)

Kirchhoff's voltage law (KVL) states that in any circuit at any instant in time, the algebraic sum of the voltages for any loop is zero. Its mathematical expression is

$$\sum v_k(t) = 0 \tag{2.2}$$

Before applying the above equation, we need to set a direction for the loop, either clockwise or counterclockwise. If the reference polarity of branch voltage is the same as the direction of the loop, a '+' is assigned for the voltage in (2.2); if the branch voltage has opposite direction of the loop, the voltage is assigned a negative sign '−'.

Fig. 2.3 shows a fraction of a circuit, which involves a loop, whose direction is clockwise. The reference polarities of the voltages are also given in the figure. According to (2.2), we have

$$v_{ab} + v_{bc} + v_{cd} + v_{de} - v_{fe} - v_{af} = 0$$

or

$$v_{ab} + v_{bc} + v_{cd} + v_{de} = v_{fe} + v_{af}$$

KVL regulates the voltages of the branches that form a loop, which is a consequence of the law of energy conservation.

KVL originally is applied to any closed loops, nonetheless it can also be applied to an open circuit or a segment of a circuit. Taking the circuit shown in Fig. 2.4 for example, in order to solve for the voltage between node a and node d, we can apply the following equation respectively

$$v_{ad} = v_{ab} + v_{bc} + v_{cd} \quad \text{or} \quad v_{ad} = v_{af} + v_{fe} + v_{ed}$$

In any circuit the voltage between any two nodes is irrelevant to the path chosen for calculation.

KCL represents the relationship of the currents of the branches connected to the same node, while KVL represents the relationship of the voltages of the branches under certain configuration of a circuit. When applying Kirchhoff's laws, the elements of the branches are of no relevance, which means Kirchhoff's laws are only related to the configuration of the circuits and have nothing to do with the elements. Therefore, KCL and KVL can be applied to any circuits, linear or nonlinear.

Discussion: Equivalent voltage/current sources

1.   Equivalent of series-connected voltage sources

How do we use four 1.5 V batteries to get a 6 V voltage?

n 个理想电压源串联，如图 2.5（a）所示，根据基尔霍夫电压定律得

$$v_{ab} = v_{s1} - v_{s2} + \cdots + v_{sn} = v_s \tag{2.3}$$

等效为一个数值为 v_s 的电压源，如图 2.5（b）所示。

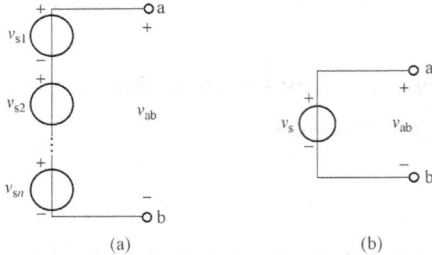

图 2.5　理想电压源串联等效
（a）原始电路；（b）等效电路
Fig. 2.5　Equivalent of series-connected voltage source
(a) Original circuit; (b) Equivalent circuit

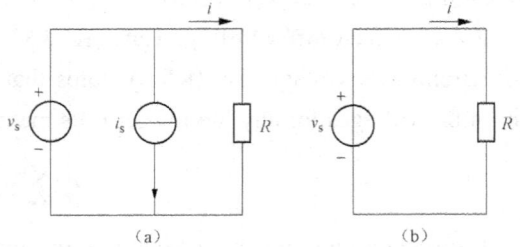

图 2.6　例 2.2 的电路图
（a）原始电路；（b）等效电路
Fig. 2.6　Circuit diagram for Example 2.2
(a) Original circuit; (b) Equivalent circuit

2. 电压源与非电压源支路的并联等效

【例 2.2】电路如图 2.6（a）所示，分析当 i_s 的大小和方向改变时，电流 i 是否会改变？

解　由于 R 两端电压已固定为 v_s，所以其电流 i 也固定，并不会受 i_s 的影响。也就是说，对 R 而言，图 2.6（a）电路的效果与图 2.6（b）电路的相同。

结论：任何一条非电压源支路与理想电压源并联后，对外的电压源特性并没有改变，等效为一个等值的理想电压源，如图 2.7 所示。

☞ 思考题 2.1：图 2.7 中等效电压源中的电流 i 等于等效前的电压源的电流 i_s 吗？

☞ 思考题 2.2：如果图 2.7 中 N 为理想电压源，且与 v_s 的数值不相等，那么并联后将出现什么情况？

3. 电流源的并联等效

n 个理想电流源并联，如图 2.8（a）所示，由基尔霍夫电流定律得

$$i = i_{s1} - i_{s2} + \cdots + i_{sn} = i_s \tag{2.4}$$

等效为一个数值为 i_s 的电流源，如图 2.8（b）所示。

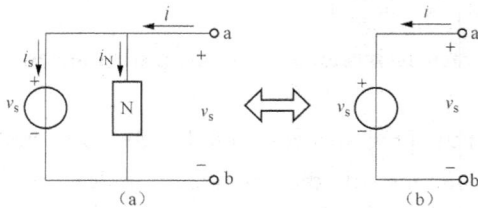

图 2.7　理想电压源与非电压源支路并联等效
（N 为除电压源外的任意元件）
（a）原始电路；（b）等效电路
Fig. 2.7　Equivalent of an ideal voltage source
and a branch containing no voltage sources
connected in parrallel
(N can be any circuit element except voltage source)
(a) Original circuit; (b) Equivalent circuit

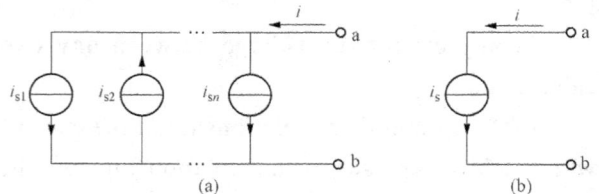

图 2.8　理想电流源并联等效
（a）原始电路；（b）等效电路
Fig. 2.8　Equivalent of parallel-connected
current sources
(a) Original circuit; (b) Equivalent circuit

This is a practical version of n ideal voltage sources connected in series. As shown in Fig. 2.5(a), the voltage between the two terminals a and b, v_{ab}, can be obtained according to KVL

$$v_{ab} = v_{s1} - v_{s2} + \cdots + v_{sn} = v_s \qquad (2.3)$$

which means the series combination of the n voltage sources is equivalent to a voltage source whose voltage takes the value of v_s, as shown in Fig. 2.5(b).

2. Equivalent of parallel combination of voltage source and other branches

Example 2.2: Consider the circuit shown in Fig. 2.6(a) When current i_s changes in value and direction, will current i change accordingly?

Solution:

As the voltage across resistor R is fixed at v_s, the current flowing through it is also fixed and will not be influenced by i_s. That is to say, for resistor R, circuits shown in Fig. 2.6(a) and Fig. 2.6(b) are equivalent.

In conclusion, if any branch that contains no voltage sources is connected in parallel with a voltage source, the voltage to the external circuit remains the same, and the two-terminal network is equivalent to the voltage source, as shown in Fig. 2.7.

✒ **Exercise 2.1**: Consider the current of the equivalent voltage source, i, in Fig. 2.7(b), and the current of the original voltage source, i_s, in Fig. 2.7(a). Are they of the same value?

✒ **Exercise 2.2**: In Fig. 2.7(a), if branch N contains only an ideal voltage source that is different from v_s, what would happen consequently?

3. Equivalent of parallel-connected current sources

Suppose we have n current sources connected in parallel, as shown in Fig. 2.8(a). Based on KCL, we have

$$i = i_{s1} - i_{s2} + \cdots + i_{sn} = i_s \qquad (2.4)$$

Apparently, the parallel combination of current sources is equivalent to a current source with the value of i_s, as shown in Fig. 2.8(b).

Gustav Robert Kirchhoff (12 March 1824 - 17 October 1887) was a German physicist who contributed to the fundamental understanding of electrical circuits and spectroscopy. Two different sets of concepts (one in circuit theory and one in spectroscopy) are named "Kirchhoff's laws" after him. Kirchhoff's laws in circuit theory, which were first described in 1845, deal with the conservation of charge and energy in electrical circuits, and are widely used in electrical engineering. Kirchhoff's three laws in spectroscopy state that a solid, liquid, or gas can be excited to emit light, and will radiate and thus produce a spectrum, which helped lead to quantum mechanics.

4. 电流源与非电流源支路的串联等效

【例 2.3】　电路如图 2.9（a）所示，分析当 v_s 的大小和方向变化时，电压 v 是否改变？

解　由于流过 R 的电流已固定为 i_s，所以 R 的电压 v 也不变，并不会受 v_s 的影响。也就是说，对 R 而言，图 2.9（a）电路的效果与图 2.9（b）的相同。

结论：任何一条非电流源支路与理想电流源串联后，对外的电流源特性并没有改变，等效为一个等值的理想电流源，如图 2.10 所示。

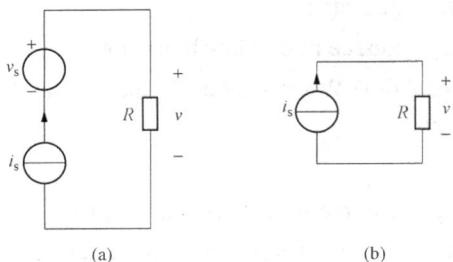

图 2.9　例 2.3 电路图

（a）原始电路；（b）等效电路

Fig. 2.9　Cirulit diagrams for Example 2.3

(a) Original circuit; (b) Equivalent circuit

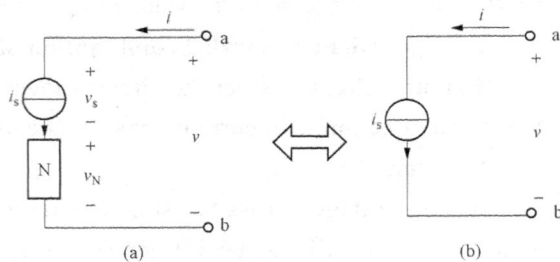

图 2.10　理想电流源与非电流源支路串联等效

（N 为除电流源外的任意元件）

（a）原始电路；（b）等效电路

Fig. 2.10　Equivalent of the series conbination of an ideal current and a current source free branch (N can be any circuit elements except current sources)

(a) Original circuit; (b) Equivalent circuit

☞　**思考题 2.3：**图 2.10 中等效电流源的电压 v 等于等效前的电流源的电压 v_s 吗？

☞　**思考题 2.4：**如果图 2.10 中 N 为理想电流源，且与 i_s 的数值不等，那么串联后将出现什么情况？

2.3　欧姆定律

如图 2.11（a）所示，电阻的电压和电流在采用关联参考方向时，任意瞬间电阻两端的电压和流过它的电流服从欧姆定律，即

$$v = Ri \tag{2.5}$$

如图 2.11（b）所示，电阻的电压和电流在采用非关联参考方向时，任意瞬间电阻两端的电压和流过它的电流服从欧姆定律，即

$$v = -Ri \tag{2.6}$$

如果电阻 R 的值不随电压或电流变化（即 R 为常数）时，则称为线性电阻；否则称为非线性电阻。如果电阻 R 的值随时间变化，称为时变电阻，否则称为时不变电阻。本书以后所讨论的电阻元件，如无特别说明，均为线性时不变电阻。线性电阻的伏安特性（欧姆定律）可用图 2.12 的曲线表示，称为线性电阻的伏安特性曲线。

电阻 R 的倒数称为电导 G

$$G = \frac{1}{R} \tag{2.7}$$

4．Equivalent of series combination of current source and other branches

Example 2.3: Consider the circuit shown in Fig. 2.9(a). When voltage v_s changes in value and polarity, will voltage v change accordingly?

Solution:

As the current flowing through resistor R is fixed at i_S, its voltage, v, remains the same and will not change as v_s changes. In other words, for resistor R, circuit shown in Fig. 2.9(a) is equivalent to the circuit of Fig. 2.9(b).

In conclusion, if a current source is connected in series to a branch that does not contain any existing current sources, the current to the external circuit is the same as that of the applied current source. The equivalent circuit is shown in Fig. 2.10 .

✈ **Exercise 2.3**:　Is the voltage v in the circuit shown in Fig. 2.10(b) the same as the voltage v_s in the circuit shown in Fig. 2.10(a)?

✈ **Exercise 2.4**:　What would happen if branch N is an ideal current source whose current is different from i_s?

2.3　Ohm's Law

Consider Fig. 2.11(a). Under a passive reference configuration, at any instance of time the voltage across the resistor and the current flowing through it is related by Ohm's law

$$v = Ri \qquad (2.5)$$

If the voltage and current are against passive reference configuration, as shown in Fig. 2.11(b), Ohm's law becomes

$$v = -Ri \qquad (2.6)$$

For a resistor whose resistance does not change with voltage or current, i.e., R is a constant, it is called a linear resistor; otherwise, it is called a nonlinear resistor. If its resistance changes with time, it is called a time-varying resistor; otherwise, we say it is a time-invariant resistor. In this book we only consider linear time-invariant resistors, if not otherwise indicated. The volt-ampere characteristic of a linear time-invariant resistor (Ohm's law) is plotted in Fig. 2.12, which is called the volt-ampere characteristic curve.

The reciprocal of resistance is called conductance, denoted by letter G

$$G = \frac{1}{R} \qquad (2.7)$$

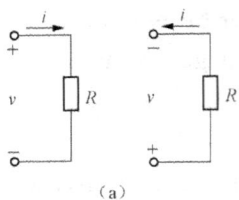
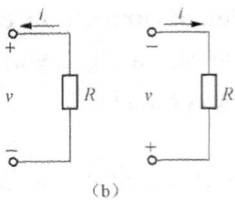
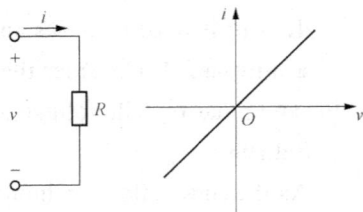

图 2.11 欧姆定律

（a）关联参考方向；（b）非关联参考方向

Fig. 2.11 Ohm's Law

(a) Passive reference configuration; (b) Against passive reference configuration

图 2.12 线性时不变电阻元件及其伏安特性

Fig. 2.12 The volt-ampere characteristic of a linear time-invariant resistor

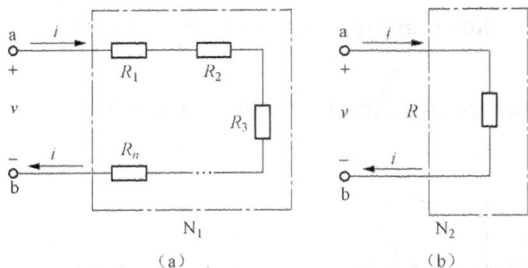

图 2.13 电阻的串联

（a）原电路；（b）等效电路

Fig. 2.13 Resistors connected in series:

(a) Original circuit; (b) Equivalent circuit

电导表示了电路元件允许电流流动的特性，电导的单位为西门子（简称西，S，$S=\Omega^{-1}$）。

2.4 电阻串并联电路的分析与化简

2.4.1 电阻的串联

两个或多个二端元件首尾相接，中间无分叉，这样的连接方式称为串联。显然，串联的每个元件中流过同一电流。考虑 n 个电阻元件串联组成的电路为 N_1，如图 2.13（a）所示。

由于每个串联电阻中流过相同的电流 i，依据基尔霍夫电压定律可以得到

$$v = R_1i + R_2i + R_3i + \cdots + R_ni = (R_1 + R_2 + R_3 + \cdots + R_n)i \tag{2.8}$$

若用图 2.13（b）所示的由单个电阻构成的电路 N_2 来代替 N_1，则必须满足 $v=Ri$ 才等效，即

$$R = R_1 + R_2 + R_3 + \cdots + R_n = \sum_{k=1}^{n} R_k \tag{2.9}$$

也就是说，这时电路 N_1 与 N_2 对外的电压电流关系（外特性）完全相同，即对外 N_1 与 N_2 是等效的。

n 个电阻 R_1，R_2，$\cdots$，R_n 串联等效为一个电阻，等效电阻 R_{eq} 等于各串联电阻之和：$R_{eq}=R_1+R_2+\cdots+R_n$。

在 n 个电阻串联的电路中，各串联电阻两端的电压为 $v_k=R_ki$（$k=1, 2, \cdots, n$），由基尔霍夫电压定律可知，$v_1+v_2+\cdots+v_n=v=Ri$。因此，每个串联电阻电压都是端口总电压的一部分，即

$$v_k = \frac{R_k}{R}v = \frac{R_k}{R_1 + R_2 + \cdots + R_n}v \tag{2.10}$$

上式称为串联电阻的电压分配公式。电阻值越大的串联电阻所分得的电压越大，在极端情况下，如果串联电路中存在一个开路元件（相当于无穷大电阻），则电路的总电压将全部分配到该开路元件上。

Conductance describes a circuit element's ability to allow the flow of current; its unit is siemens (abbreviated S and $S=\Omega^{-1}$).

2.4 Analysis and Simplification of Resistor Circuits

2.4.1 Resistors in series

When two or more two-terminal elements connected end-to-end and no other path is connected in the middle, we say that they are connected in series. Apparently, the currents flowing through the elements are the same. Consider circuit N_1 where n resistors are connected in series, as shown in Fig. 2.13(a). As the same current, i, flows through all the resistors, we can obtain the following equation according to KVL

$$v = R_1 i + R_2 i + R_3 i + \cdots + R_n i = (R_1 + R_2 + R_3 + \cdots + R_n)i \tag{2.8}$$

If we substitute circuit N_1 by a single resistor, as circuit N_2 shown in Fig. 2.13(b), the constraint of $v=Ri$ must be satisfied. Therefore, we have

$$R = R_1 + R_2 + R_3 + \cdots + R_n = \sum_{k=1}^{n} R_k \tag{2.9}$$

In this case, circuits N_1 and N_2 share the same volt-ampere characteristic to external circuit, i.e., circuits N_1 and N_2 are equivalent.

The series connection of n resistors, R_1, R_2, $\cdots$, R_n, is equivalent to a single resistor, R_{eq}, the resistance of which equals to the sum of the n resistances:

$$R_{eq}=R_1+R_2+\cdots+R_n$$

In a circuit where n resistors are connected in series, each resistor has a voltage of

$$v_k=R_k i \ (k=1,2,\cdots, n)$$

According to KVL, we have

$$v_1+v_2+\cdots+v_n=v=Ri$$

Therefore, the voltage of each resistor is a fraction of the total voltage at the terminal

$$v_k = \frac{R_k}{R} v = \frac{R_k}{R_1 + R_2 + \cdots + R_n} v \tag{2.10}$$

This is known as the voltage division principle. A larger resistance bears larger voltage. In extreme condition where an open circuit element (equivalent to an infinite resistance) is connected in series, the voltage is totally applied to this open circuit element.

【**例 2.4**】　如图 2.14 所示的空载分压器电路，设输入的信号电压 v_i 为 50 V，$R_1=1\ \text{k}\Omega$，$R_2= 9\ \text{k}\Omega$，问从 ab 端得到的输出电压 v_o 为多少？

解　输出电压即为电阻 R_1 上的分压，由式（2.10），可得

$$v_o = \frac{R_1}{R_1 + R_2} v_i = 50 \times \frac{1}{9 + 1} = 5\ \text{V}$$

如果将图 2.14 中的两个电阻合为一个电阻 R，即 $R=R_1+R_2$。并在 R 上设一个可以滑动的触点，滑动触点相当于改变 R_1 和 R_2 的比例，而保持 $R=R_1+R_2$ 不变，如图 2.15 所示。输出电压与电阻 R_1 呈正比地变化。调节接触点我们可以得到一个大小从 0 到 v_i 连续可变而极性不变的电压。这种带有中间滑动端的电阻元件称为电位器。收音机中就是用电位器来调节音量（音频输出电压）的大小的。

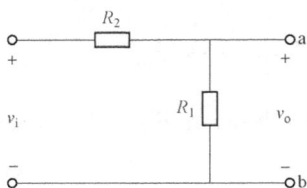

图 2.14　空载分压器电路　　　　　　　图 2.15　电位器的电路图
Fig. 2.14　A voltage divider circuit　　　Fig. 2.15　A circuit diagram for potentiometer

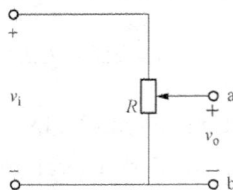

2.4.2　电阻的并联

两个或多个二端元件连接在一对节点之间，这样的连接方法称为并联。显然，并联的每个元件具有相同的电压。考虑 n 个电阻元件 $R_1, R_1, \cdots, R_n$ 并联而成的电路 N_1，如图 2.16（a）所示，利用基尔霍夫电流定律和欧姆定律，可以得到

$$i = i_1 + i_2 + \cdots + i_n = \frac{v}{R_1} + \frac{v}{R_2} + \cdots + \frac{v}{R_n} = v\left(\frac{1}{R_1} + \frac{1}{R_2} + \cdots + \frac{1}{R_n} \right) = v\sum_{k=1}^{n} \frac{1}{R_k} \qquad (2.11)$$

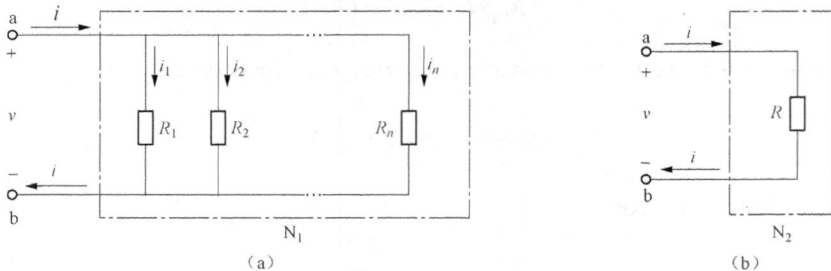

图 2.16　电阻的并联
（a）原电路；（b）等效电路
Fig. 2.16　Resistors connected in parallel
(a) Original circuit; (b) Equivalent circuit

若用图 2.16（b）所示的由单一电阻 R 构成的电路 N_2 来等效替代 N_1，则必须满足

$$i = v\frac{1}{R} \qquad (2.12)$$

即

$$\frac{1}{R} = \frac{1}{R_1} + \frac{1}{R_2} + \cdots + \frac{1}{R_n} = \sum_{k=1}^{n} \frac{1}{R_k} \qquad (2.13)$$

Example 2.4 Fig. 2.14 shows a voltage divider circuit. The input voltage is $v_i=50$ V, the resistances are $R_1=1$ kΩ and $R_2=9$ kΩ, respectively. Find the output voltage v_o.

Solution:

The output voltage v_o is the input voltage times the ratio of R_1 to the total resistance

$$v_o = \frac{R_1}{R_1 + R_2} v_i = 50 \times \frac{1}{9+1} = 5V$$

Suppose we combine the two resistors in Fig. 2.14 into one resistor R, i.e. $R=R_1+R_2$. Suppose further that we have a contact that can move along resistor R. As the contact moves, the ratio between R_1 and R_2 changes accordingly, while their sum remains the same, as shown in Fig. 2.15. In this case, the output voltage changes proportion ally with R_1. Therefore, by adjusting the position of the contact, we can get an output voltage varying continuously from 0 to v_i. Such a resistor with a contact is known as a potentiometer, which is commonly used to adjust the volume (the voltage of audio output) of a radio.

2.4.2 Resistors in parallel

We say that two or more circuit elements are connected in parallel if both the ends of one element are connected directly to the corresponding ends of another element. Apparently, in this case all the elements have the same voltage. Consider the circuit shown in Fig. 2.16(a), where n resistors R_1, R_2, $\cdots$, R_n are connected in parallel to form a circuit N_1. Based on KCL and Ohm's law, we have

$$i = i_1 + i_2 + \cdots + i_n = \frac{v}{R_1} + \frac{v}{R_2} + \cdots + \frac{v}{R_n} = v\left(\frac{1}{R_1} + \frac{1}{R_2} + \cdots + \frac{1}{R_n}\right) = v\sum_{k=1}^{n}\frac{1}{R_k} \qquad (2.11)$$

To substitute circuit N_1 by circuit N_2 that contains only one resistor, the following constraint must be sufficed

$$i = v\frac{1}{R} \qquad (2.12)$$

i.e.

$$\frac{1}{R} = \frac{1}{R_1} + \frac{1}{R_2} + \cdots + \frac{1}{R_n} = \sum_{k=1}^{n}\frac{1}{R_k} \qquad (2.13)$$

这时电路 N_1 与 N_2 对外的电压电流关系（外特性）完全相同，也就是说，N_1 与 N_2 等效。

> n 个电阻 $R_1, R_2, \cdots, R_n$ 并联等效为一个电阻，等效电阻 R_{eq} 的倒数等于各并联电阻倒数之和，即
>
> $$\frac{1}{R_{eq}} = \frac{1}{R_1} + \frac{1}{R_2} + \cdots + \frac{1}{R_n}$$

如果我们采用电导 G 表示电阻元件的参数，则

> n 个电导 $G_1, G_2, \cdots, G_n$ 并联等效为一个电导，等效电导 G_{eq} 等于各并联电导之和
> $$G_{eq} = G_1 + G_2 + \cdots + G_n$$

在 n 个电阻并联的电路中，流过各并联电阻的电流为 $i_k = \dfrac{v}{R_k}$（$k = 1, 2, \cdots, n$），由基尔霍夫电流定律可知，$i_1 + i_2 + \cdots + i_n = i = \dfrac{v}{R}$。因此，每个并联电阻的电流都是电路总电流的一部分，即

$$i_k = \frac{R}{R_k} i = \frac{\frac{1}{R_k}}{\sum\limits_{m=1}^{n} \frac{1}{R_m}} i = \frac{G_k}{\sum\limits_{m=1}^{n} G_m} i \tag{2.14}$$

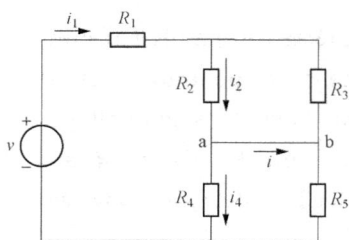

图 2.17　例 2.5 的电路图

Fig. 2.17　Circuit diagram for Example 2.5

式（2.14）称为并联电阻的电流分配公式。电阻值越小的并联电阻所分得的电流越大。在极端情况下，如果并联电路中存在一个短路元件（相当于电阻为零），则电路的总电流将全部分配给该短路元件，这种情况在电子技术中称为旁路。

【例 2.5】电路如图 2.17 所示，已知 $R_1 = 1\ \Omega$，$R_2 = 3\ \Omega$，$R_3 = 6\ \Omega$，$R_4 = 12\ \Omega$，$R_5 = 6\ \Omega$，$v = 21\ V$，求电路中的电流 i。

解　分析该电路可知，R_2 与 R_3、R_4 与 R_5 分别并联，然后再与 R_1 串联。应用串、并联等效，得电流

$$i_1 = \frac{v}{R_1 + R_2 /\!/ R_3 + R_4 /\!/ R_5} = \frac{21}{1 + \dfrac{3 \times 6}{3 + 6} + \dfrac{12 \times 6}{12 + 6}} = 3\ A$$

再应用式（2.14）得

$$i_2 = \frac{R_3}{R_2 + R_3} i_1 = \frac{6}{3 + 6} \times 3 = 2\ A$$

$$i_4 = \frac{R_5}{R_4 + R_5} i_1 = \frac{6}{12 + 6} \times 3 = 1\ A$$

对节点 a 应用基尔霍夫电流定律，得电流

$$i = i_2 - i_4 = 2 - 1 = 1\ A$$

Under this condition, circuits N_1 and N_2 share the same volt-ampere characteristic to external circuit, i.e., circuits N_1 and N_2 are equivalent.

The parallel connection of n resistances $R_1, R_2, \cdots, R_n$ equals to an equivalent resistance R_{eq}, the reciprocal of which equals to the sum of the reciprocal of each resistance

$$\frac{1}{R_{eq}} = \frac{1}{R_1} + \frac{1}{R_2} + \cdots + \frac{1}{R_n}$$

It is often more convenient to use conductance rather than resistance when dealing with resistors in parallel.

For the parallel connection of n conductances $G_1, G_2, \cdots, G_n$, its equivalent conductance G_{eq} equals to the sum of their individual conductance

$$G_{eq} = G_1 + G_2 + \cdots + G_n$$

In a circuit where n resistors are connected in parallel, the current flowing through each resistor is $i_k = \frac{v}{R_k}$ ($k = 1, 2, \cdots, n$). According to KCL, we have $i_1 + i_2 + \cdots + i_n = i = \frac{v}{R}$. Hence, the current of each resistor is a fraction of the total current

$$i_k = \frac{R}{R_k} i = \frac{\frac{1}{R_k}}{\sum_{m=1}^{n} \frac{1}{R_m}} i = \frac{G_k}{\sum_{m=1}^{n} G_m} i \qquad (2.14)$$

This is known as the current division principle. A smaller resistance bears larger current. In extreme case where a short circuit element (whose resistance is 0) is connected in parallel with other circuit elements, the current flows entirely through the short circuit element. This is called a bypass in electrical circuits.

Example 2.5: Solve for the circuit shown in Fig. 2.17, where $R_1=1\ \Omega$, $R_2=3\ \Omega$, $R_3=6\ \Omega$, $R_4=12\ \Omega$, $R_5=6\ \Omega$, and $v=21$ V. Find current i.

Solution:

In this circuit, R_2 and R_3 are connected in parallel, R_4 and R_5 are connected in parallel, then they are connected to R_1 in series. According to the equivalent resistance of series and parallel connected resistors, we have

$$i_1 = \frac{v}{R_1 + R_2 \text{//} R_3 + R_4 \text{//} R_5} = \frac{21}{1 + \dfrac{3 \times 6}{3+6} + \dfrac{12 \times 6}{12+6}} = 3\ \text{A}$$

Then, according to (2.14), we get

$$i_2 = \frac{R_3}{R_2 + R_3} i_1 = \frac{6}{3+6} \times 3 = 2\ \text{A}$$

$$i_4 = \frac{R_5}{R_4 + R_5} i_1 = \frac{6}{12+6} \times 3 = 1\ \text{A}$$

Applying KCL at node a, we obtain

$$i = i_2 - i_4 = 2 - 1 = 1\ \text{A}$$

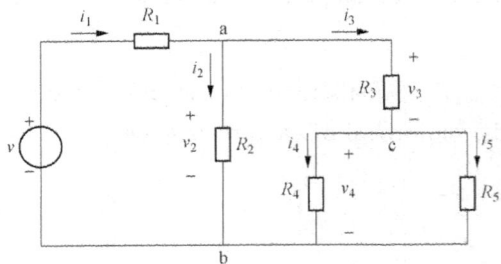

图 2.18　例 2.6 的电路图

Fig. 2.18　Circuit diagram for Example 2.6

2.4.3　电阻的串并联

电阻串联的特点：

（1）流过每个电阻的电流相等；

（2）各串联电阻对总电压进行分压；

（3）等效电阻大于任何一个串联电阻。

电阻并联的特点：

（1）所有电阻两端的电压相等；

（2）各并联电阻对总电流进行分流；

（3）等效电阻小于任何一个并联电阻。

【例 2.6】 电路如图 2.18 所示。已知 $R_1=5\ \Omega$，$R_2=18\ \Omega$，$R_3=6\ \Omega$，$R_4=4\ \Omega$，$R_5=12\ \Omega$，$v=165$ V。求电路中标出的各电流、电压。

解　电阻 R_4 和 R_5 并联，其等效电阻为

$$R_{cb} = R_4 \mathbin{/\mkern-5mu/} R_5 = \frac{4 \times 12}{4 + 12} = 3\ \Omega$$

支路 acb 的等效电阻为

$$R_{acb} = R_3 + R_{cb} = 6 + 3 = 9\ \Omega$$

节点 a 和 b 之间的等效电阻 R_{ab} 为电阻 R_2 和 R_{acb} 的并联

$$R_{ab} = R_2 \mathbin{/\mkern-5mu/} R_{acb} = \frac{18 \times 9}{18 + 9} = 6\ \Omega$$

电路的总等效电阻 R_{eq} 为 R_1 和 R_{ab} 的串联

$$R_{eq} = R_1 + R_{ab} = 5 + 6 = 11\ \Omega$$

根据欧姆定律，总电流为

$$i_1 = \frac{v}{R_{eq}} = \frac{165}{11} = 15\ \text{A}$$

根据分压原理，有

$$v_2 = \frac{R_{ab}}{R_{eq}} v = \frac{6}{11} \times 165 = 90\ \text{V}$$

为电阻 R_2 应用欧姆定律，可得

$$i_2 = \frac{v_2}{R_2} = \frac{90}{18} = 5\ \text{A}$$

为等效电阻 R_{acb} 应用欧姆定律，可得

$$i_3 = \frac{v_2}{R_{acb}} = \frac{90}{9} = 10\ \text{A}$$

应用分流原理，可得

$$i_4 = \frac{R_5}{R_4 + R_5} i_3 = \frac{12}{4 + 12} \times 10 = 7.5\ \text{A}$$

$$i_5 = \frac{R_4}{R_4 + R_5} i_3 = \frac{4}{4 + 12} \times 10 = 2.5\ \text{A}$$

2.4.3   Resistors in series and parallel

For resistors connected in series, we know that

（1）The current flowing through the resistors is the same;

（2）Each resistor shares a fraction of the total voltage;

（3）The equivalent resistance is larger than any individual resistance.

For resistors connected in parallel, we can conclude that

（1）The voltage across the resistors is the same;

（2）Each resistor shares a fraction of the total current;

（3）The equivalent resistance is smaller than any individual resistance.

Example 2.6: Find the currents and voltages labeled in Fig. 2.18, where R_1=5 Ω, R_2=18 Ω, R_3=6 Ω, R_4=4 Ω, R_5=12 Ω, and v =165 V.

Solution:

R_4 and R_5 are connected in parallel and the equivalent resistance is

$$R_{cb} = R_4 \; // \; R_5 = \frac{4 \times 12}{4 + 12} = 3 \; \Omega$$

The equivalent resistance in branch a-c-b is

$$R_{acb} = R_3 + R_{cb} = 6 + 3 = 9 \; \Omega$$

The equivalent resistance between node a and node b is the parallel connection of R_2 and R_{acb}

$$R_{ab} = R_2 \; // \; R_{acb} = \frac{18 \times 9}{18 + 9} = 6 \; \Omega$$

The total equivalent resistance is the series connection of R_1 and R_{ab}

$$R_{eq} = R_1 + R_{ab} = 5 + 6 = 11 \; \Omega$$

According to Ohm's law, the total current, i_1, can be calculated from

$$i_1 = \frac{v}{R_{eq}} = \frac{165}{11} = 15 \; A$$

According to voltage division principle, we have

$$v_2 = \frac{R_{ab}}{R_{eq}} v = \frac{6}{11} \times 165 = 90 \; V$$

Applying Ohm's law to R_2, the current of i_2 can be obtained

$$i_2 = \frac{v_2}{R_2} = \frac{90}{18} = 5 \; A$$

Applying Ohm's law to equivalent resistance R_{acb}, we have

$$i_3 = \frac{v_2}{R_{acb}} = \frac{90}{9} = 10 \; A$$

According to the current division principle, we have

$$i_4 = \frac{R_5}{R_4 + R_5} i_3 = \frac{12}{4 + 12} \times 10 = 7.5 \; A$$

$$i_5 = \frac{R_4}{R_4 + R_5} i_3 = \frac{4}{4 + 12} \times 10 = 2.5 \; A$$

再次应用欧姆定律计算电阻 R_3 和 R_4 两端电压，分别为

$$v_3 = R_3 i_3 = 6 \times 10 = 60 \ \text{V}$$

$$v_4 = R_4 i_4 = 4 \times 7.5 = 30 \ \text{V}$$

2.5　电阻 Y 连接和△连接的等效变换

在某些情况下，电阻的连接方式既非串联也非并联，如图 2.19 所示。其中图 2.19（a）所示的连接方式称为 Y 连接，图 2.19（b）所示的连接方式称为△连接。这两种连接方式下，三个电阻都通过节点 a、b、c 分别与外电路连接。当两组电阻的阻值满足一定条件时，这两种连接方式相对于外电路而言是等效的，即电阻 Y 连接和△连接的等效变换。此时需满足的条件为各节点之间的电压相同，即

$$v_{ab} = \tilde{v}_{ab}, v_{bc} = \tilde{v}_{bc}, v_{ca} = \tilde{v}_{ca} \tag{2.15}$$

以及流入各节点的电流相同，即

$$i_a = \tilde{i}_a, i_b = \tilde{i}_b, i_c = \tilde{i}_c \tag{2.16}$$

对于 Y 连接的电路，我们可根据基尔霍夫电流和电压定律列出下列方程

$$i_a + i_b + i_c = 0 \tag{2.17}$$

$$v_{ab} = R_a i_a - R_b i_b \tag{2.18}$$

$$v_{bc} = R_b i_b - R_c i_c \tag{2.19}$$

$$v_{ca} = R_c i_c - R_a i_a \tag{2.20}$$

对于△连接的电路，我们可为各节点列基尔霍夫电流定律方程

$$\tilde{i}_a = \frac{\tilde{v}_{ab}}{R_{ab}} - \frac{\tilde{v}_{ca}}{R_{ca}} \tag{2.21}$$

$$\tilde{i}_b = \frac{\tilde{v}_{bc}}{R_{bc}} - \frac{\tilde{v}_{ab}}{R_{ab}} \tag{2.22}$$

$$\tilde{i}_c = \frac{\tilde{v}_{ca}}{R_{ca}} - \frac{\tilde{v}_{bc}}{R_{bc}} \tag{2.23}$$

联立式（2.17）、式（2.18）～式（2.20）中的任意两个、式（2.21）～式（2.23）6 个方程，可以得到如下两组关于电阻阻值的关系式

$$\begin{cases} R_{ab} = \dfrac{R_a R_b + R_b R_c + R_c R_a}{R_c} \\[3mm] R_{bc} = \dfrac{R_a R_b + R_b R_c + R_c R_a}{R_a} \\[3mm] R_{ca} = \dfrac{R_a R_b + R_b R_c + R_c R_a}{R_b} \end{cases} \tag{2.24}$$

Again, using Ohm's law to find the voltages across R_3 and R_4, respectively

$$v_3 = R_3 i_3 = 6 \times 10 = 60 \text{ V}$$

$$v_4 = R_4 i_4 = 4 \times 7.5 = 30 \text{ V}$$

2.5 Transformations between Wye–connected and Delta–connected Resistors

There are cases when resistors are connected neither in series nor in parallel, as shown in Fig. 2.19, in which the circuit of Fig. 2.19(a) is called wye connection and the one given in Fig. 2.19(b) is called delta connection. In both cases, the three resistors are connected to the external circuit through nodes a, b and c, respectively. If the resistances of the two groups of resistors satisfy certain conditions, the two circuits are equivalent to the external circuit. This is the equivalent transformation between wye-connected and delta-connected resistors. The following conditions should be met

The voltage across each pair of nodes should be the same

$$v_{ab} = \tilde{v}_{ab}, v_{bc} = \tilde{v}_{bc}, v_{ca} = \tilde{v}_{ca} \tag{2.15}$$

The current entering each node should be the same as well

$$i_a = \tilde{i}_a, i_b = \tilde{i}_b, i_c = \tilde{i}_c \tag{2.16}$$

If we apply KCL and KVL to the wye-connected circuit, we have

$$i_a + i_b + i_c = 0 \tag{2.17}$$

$$v_{ab} = R_a i_a - R_b i_b \tag{2.18}$$

$$v_{bc} = R_b i_b - R_c i_c \tag{2.19}$$

$$v_{ca} = R_c i_c - R_a i_a \tag{2.20}$$

For the delta-connected circuit, the following KCL equations can be written for each node

$$\tilde{i}_a = \frac{\tilde{v}_{ab}}{R_{ab}} - \frac{\tilde{v}_{ca}}{R_{ca}} \tag{2.21}$$

$$\tilde{i}_b = \frac{\tilde{v}_{bc}}{R_{bc}} - \frac{\tilde{v}_{ab}}{R_{ab}} \tag{2.22}$$

$$\tilde{i}_c = \frac{\tilde{v}_{ca}}{R_{ca}} - \frac{\tilde{v}_{bc}}{R_{bc}} \tag{2.23}$$

Solving the six equations of (2.17), any two of (2.18) $\sim$ (2.20), and (2.21) $\sim$ (2.23) simultaneously, we can write the relationship between the two groups of resistances as follows

$$\begin{cases} R_{ab} = \dfrac{R_a R_b + R_b R_c + R_c R_a}{R_c} \\[2ex] R_{bc} = \dfrac{R_a R_b + R_b R_c + R_c R_a}{R_a} \\[2ex] R_{ca} = \dfrac{R_a R_b + R_b R_c + R_c R_a}{R_b} \end{cases} \tag{2.24}$$

$$\begin{cases} R_\text{a} = \dfrac{R_\text{ab}R_\text{ca}}{R_\text{ab} + R_\text{bc} + R_\text{ca}} \\[2ex] R_\text{b} = \dfrac{R_\text{bc}R_\text{ab}}{R_\text{ab} + R_\text{bc} + R_\text{ca}} \\[2ex] R_\text{c} = \dfrac{R_\text{ca}R_\text{bc}}{R_\text{ab} + R_\text{bc} + R_\text{ca}} \end{cases} \tag{2.25}$$

式（2.24）即为根据 Y 连接的电阻确定等效△连接电阻的公式，而式（2.25）为根据△连接电阻确定等效 Y 连接电阻的公式。当三个电阻阻值相同时，上述关系简化为

$$R_\triangle = 3R_\text{Y} \tag{2.26}$$

$$R_\text{Y} = R_\triangle / 3 \tag{2.27}$$

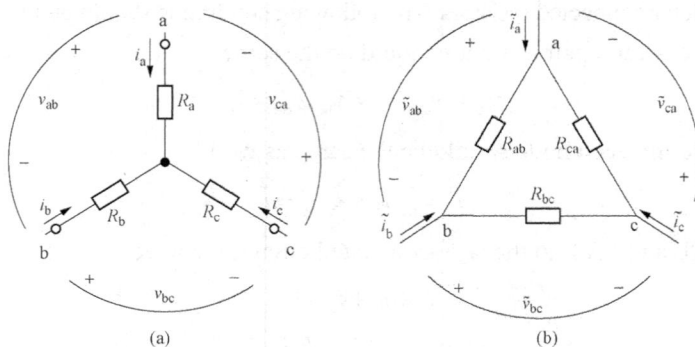

图 2.19　电阻的 Y 连接及△连接

（a）电阻的 Y 连接；（b）电阻的△连接

Fig. 2.19　Y-connected resistors and △-connected resistors

(a) Y-connected vesistors; (b) △-connected resistors

$$\begin{cases} R_a = \dfrac{R_{ab}R_{ca}}{R_{ab}+R_{bc}+R_{ca}} \\[2mm] R_b = \dfrac{R_{bc}R_{ab}}{R_{ab}+R_{bc}+R_{ca}} \\[2mm] R_c = \dfrac{R_{ca}R_{bc}}{R_{ab}+R_{bc}+R_{ca}} \end{cases} \tag{2.25}$$

We can apply (2.24) to determine the resistances of delta-connected resistors according to the wye-connected resistances, and use (2.25) to calculate the resistances of wye-connected resistors based on the delta-connected resistors. If the three resistances are of the same value, the two equations are simplified as

$$R_\triangle = 3R_Y \tag{2.26}$$

$$R_Y = R_\triangle/3 \tag{2.27}$$

▶ Problems

P2.1 For the circuit shown in Fig. P2.1, where $i_1 = -1\,\text{A}$ and $i_3 = 2\,\text{A}$, find current i_2.

P2.2 For the circuit shown in Fig. P2.2, where $v_s = 10$ V, $i_1 = -4\,\text{A}$ and $i_2 = 1$ A, find current i_3.

P2.3 For the circuit shown in Fig. P2.3, find current i.

P2.4 For the circuit shown in Fig. P2.4, find voltage v_{ab}.

P2.5 For the circuit shown in Fig. P2.5, where $i_s = 5\,\text{A}$, $R_1 = 1$ Ω and $R_2 = 2$ Ω, find voltage v_2. The value of R_1 is adjustable. Will the value of v_2 change as we adjust the value of R_1?

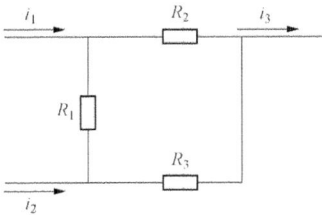

Fig. P2.1

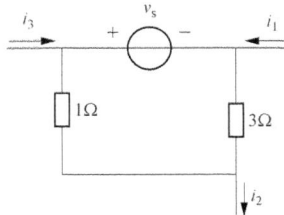

Fig. P2.2

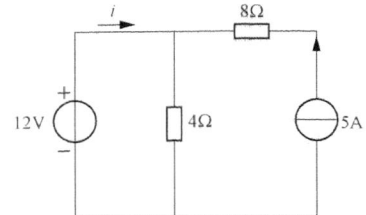

Fig. P2.3

P2.6 For the two circuits shown in Fig. P2.6, calculate i_1 and i_2, respectively. Explain that the two circuits are equivalent to i_1 and i_2.

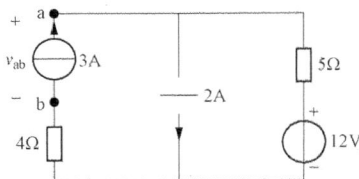

Fig. P2.4

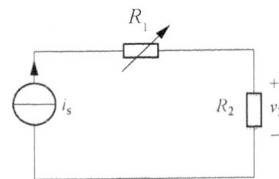

Fig. P2.5

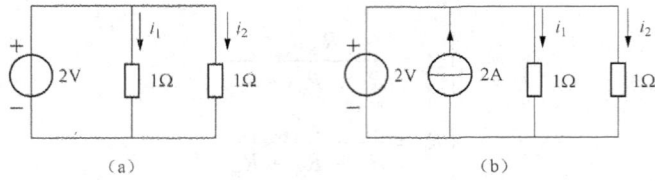

Fig. P2.6

P2.7　All the resistances in Fig. P2.7 are of $8\ \Omega$. Calculate the equivalent resistance R_{ab}.

P2.8　Find the equivalent resistance R_{ab} of Fig. P2.8.

P2.9　For the circuit shown in Fig. P2.9, where $v_s = 24$ V, $R_1 = 20\ \Omega$, $R_2 = 30\ \Omega$, $R_3 = 15\ \Omega$, $R_4 = 100\ \Omega$, $R_5 = 25\ \Omega$, $R_6 = 8\ \Omega$. Solve for the power of R_6.

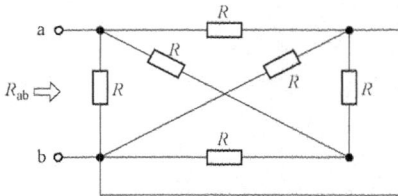

Fig. P2.7

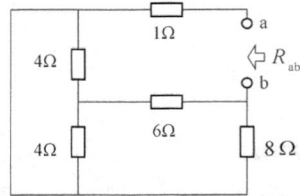

Fig. P2.8

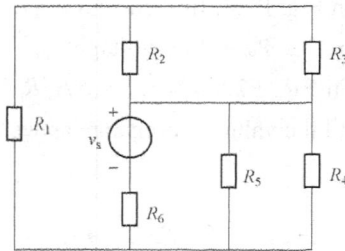

Fig. P2.9

第3章
Chapter 3

经典电路分析方法
Circuit Analysis Methods

本章介绍几种常用的电路分析方法，着重电路分析的技巧，这对整个课程的学习和电路问题的处理都是有益的。

3.1 节点电压法

如果用支路电压作为基本变量，对电路进行分析，那么，只要确定了各支路电压，支路电流等完全可以确定。

电路中的电压实际就是单位电荷在两个节点之间所转移的能量。如果将单位电荷在某节点处的能量定义为电位，则每条支路的电压实际上就是该支路所连两个节点的电位差。因此，如果知道了电路中各个节点的电位，则可由节点电位求解所有支路的电压和电流，即以节点电位作为分析电路的基本变量，所以节点电压法也称节点电位法。

3.1.1 节点电压的概念

在电路中选择一个节点作为参考节点，设其电位为零，在电路图上用"⏚"标记。那么，所有其他节点的电位实际上就是该节点与参考节点之间的电压，我们将其定义为非参考节点的节点电压。节点电压的参考极性是以参考节点为负、非参考节点为正。

值得注意的是，节点电压只有在选定了参考节点后才有意义。因此，采用节点电压法分析电路，必须首先在电路中选定参考节点。

3.1.2 节点电压方程

现以图 3.1（a）为例来说明以节点电压为基本变量的电路方程组形式。

在图 3.1（a）所示电路中，共有三个节点，选定一个节点作为参考节点后，只有两个独立节点（即非参考节点）a、b。对于每个独立节点，列写 KCL 方程

节点 a：
$$i_3 = i_1 + i_2$$

节点 b：
$$i_3 = i_4 + i_5$$

将 a、b 两节点处的电位分别表示为 u_a 和 u_b。根据 KVL 和欧姆定律，进行变量代换，并将各个支路电流用节点电压表示

$$i_1 = \frac{v_1 - u_a}{R_1}；\quad i_2 = \frac{v_2 - u_a}{R_2}；\quad i_3 = \frac{u_a - u_b}{R_3}；\quad i_4 = \frac{u_b}{R_4}；\quad i_5 = \frac{u_b + v_5}{R_5}$$

将上述各支路电流代入 a、b 两节点电流方程，整理得

$$\frac{u_a}{R_1} + \frac{u_a}{R_2} + \frac{u_a - u_b}{R_3} = \frac{v_1}{R_1} + \frac{v_2}{R_2} \tag{3.1}$$

$$\frac{u_b - u_a}{R_3} + \frac{u_b}{R_4} + \frac{u_b}{R_5} = -\frac{v_5}{R_5} \tag{3.2}$$

从上面推导过程可以看出，节点电压方程是由节点电压表示的 KCL 方程，因此，KCL 是节点电压分析法的基本出发点。

In this chapter, we will introduce a number of circuit analysis methods which are common practice. We will focus on the circuit analysis skills, which is of great importance to this course and to solving circuit problems.

3.1 Node Voltage Method

If we use the voltage across a branch as the variable to analyze a circuit, the branch currents can be calculated as the corresponding branch voltages are obtained.

As defined in Chapter 1, voltage between two nodes of a circuit is the energy transferred per unit of charge that flows from one node to the other. If the energy carried by per unit of charge at a node is defined as potential, the branch voltage is the difference of the potentials between the two nodes of the branch. Therefore, the branch voltages and branch currents of the branch can be calculated from node potentials, which are the basic variables used in circuit analysis. This is why sometimes node voltage method is also referred to as node potential method.

3.1.1 Definition of node voltage

Select a node from the circuit as the reference node, whose potential is zero. In a circuit diagram, the reference node is labeled by the symbol of $\perp$. Therefore, the potentials of the rest nodes are the voltages between the nodes and the reference node, respectively. As the reference node has zero potential, the potentials of the rest nodes are defined as node voltages. The reference polarity of a node voltage is positive at this node and negative at the reference node.

We must pay attention that we can only use node voltage when the reference node is determined. Therefore, the first step of the node voltage method is to select a proper reference node.

3.1.2 Node voltage equations

We use the circuit shown in Fig. 3.1(a) to illustrate how to set up node voltage equations. In this circuit there are three nodes. We select the bottom one as the reference node, and consequently, the two top ones are the independent nodes (non-reference nodes), denoted by node a and node b. For each independent node, we can write the following KCL equations.

For node a $\qquad\qquad i_3 = i_1 + i_2$

For node b $\qquad\qquad i_3 = i_4 + i_5$

Denote the potentials at nodes a and b by u_a and u_b, respectively. According to KVL and Ohm's law, each branch current can be calculated from

$$i_1 = \frac{v_1 - u_a}{R_1}, \quad i_2 = \frac{v_2 - u_a}{R_2}, \quad i_3 = \frac{u_a - u_b}{R_3}, \quad i_4 = \frac{u_b}{R_4}, \quad i_5 = \frac{u_b + v_5}{R_5}$$

Substitute the above currents into the KCL equations of nodes A and B, respectively, we have

$$\frac{u_a}{R_1} + \frac{u_a}{R_2} + \frac{u_a - u_b}{R_3} = \frac{v_1}{R_1} + \frac{v_2}{R_2} \tag{3.1}$$

$$\frac{u_b - u_a}{R_3} + \frac{u_b}{R_4} + \frac{u_b}{R_5} = -\frac{v_5}{R_5} \tag{3.2}$$

From the above derivation it can be seen that node voltage equations are KCL equations expressed by node voltages. Therefore, the node voltage method is rooted from KCL.

仔细观察上面的节点电压方程，我们不难看出，如果将电路中所有电压源模型（带有串联电阻的电压源）全部转换为电流源模型，如图 3.1（b）所示，则节点电压方程具有下面结构：方程的左边为流出节点的电阻支路电流总和，而方程的右边为注入节点的纯电流源支路电流的总和，即

$$\sum_{\text{流出节点}} i_{Rk} = \sum_{\text{流入节点}} i_{Sk} \tag{3.3}$$

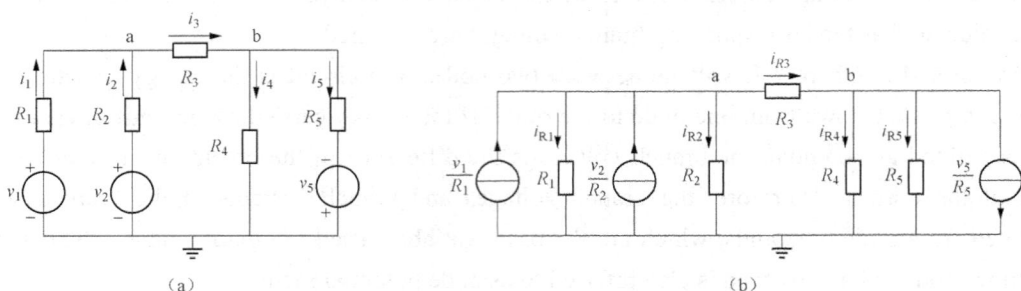

图 3.1　节点电压方程的推导

（a）原始电路；（b）将电压源等效为电流源之后的等效电路

Fig. 3.1　Derivation of node voltage equations

(a) Original circuit; (b) Equivalent circuit when voltage sources are substituted by current sources

3.1.3　由观察法快速建立节点电压方程

我们将节点电压方程作进一步整理，将每个节点电压变量的系数合并，得到

节点 a：
$$\left(\frac{1}{R_1} + \frac{1}{R_2} + \frac{1}{R_3}\right)u_a + \left(-\frac{1}{R_3}\right)u_b = \frac{v_1}{R_1} + \frac{v_2}{R_2}$$

节点 b：
$$\left(-\frac{1}{R_3}\right)u_a + \left(\frac{1}{R_4} + \frac{1}{R_5} + \frac{1}{R_3}\right)u_b = -\frac{v_5}{R_5}$$

在节点 a 的方程左边，节点 a 电压 u_a 的系数为所有连接到节点 a 的电阻支路电导之和，称为节点 a 的自电导，记作 G_{aa}；节点 b 电压 u_b 的系数为连接在节点 a 和 b 之间的电阻支路电导之和的负值，称为节点 a 与节点 b 之间的互电导，记做 G_{ab}。如果两个节点之间没有电阻支路直接相连，则相应的互电导为零；显然，$G_{ab}=G_{ba}$。

节点 A 的方程右边的常数项为连接到节点 a 的纯电流源支路流入节点 A 的电源电流之代数和，记做 i_{sa}；如果没有纯电流源支路接到节点 a，则 $i_{sa}=0$。

节点 B 的方程具有完全相同的结构，因此，我们可以从电路的组成结构直接根据节点电压方程的结构特点，通过观察直接列写电路方程，无须一步一步从电路基本定律去推导节点电压方程。

综上所述，节点电压方程的一般形式为

$$G_{kk}u_k + \sum_{j \neq k} G_{kj}u_j = i_{sk} \tag{3.4}$$

需要注意，采用节点电压法分析电路，电路中电源宜取电流源模型，如果电路中含有电压源（带串联电阻），可利用电源模型的转换使之转化为电流源模型，如图 3.1（b）所示。

【例 3.1】　在图 3.2 所示电路中设 d 为参考节点。试列出各节点电压方程。

解　该电路有三个独立节点 a、b、c。我们采用观察法直接列写节点电压方程

Examine the above node voltage equations we can find that if all the voltage sources (the voltage source connected in series with a resistor) are substituted by current sources, as shown in Fig. 3.1(b), node voltage equations have the following format: the left hand side of the equation is the sum of the currents of resistor branches leaving the node, while the right hand side of the equation is the sum of the currents of current sources entering the node. That is

$$\sum_{\text{leaving the node}} i_{Rk} = \sum_{\text{entering the node}} i_{sk} \tag{3.3}$$

3.1.3 Another version of node voltage equations

We can further sort (3.1) and (3.2) by combining the coefficients of each node voltage.

For node a
$$\left(\frac{1}{R_1} + \frac{1}{R_2} + \frac{1}{R_3}\right)u_a + \left(-\frac{1}{R_3}\right)u_b = \frac{v_1}{R_1} + \frac{v_2}{R_2}$$

For node b
$$\left(-\frac{1}{R_3}\right)u_a + \left(\frac{1}{R_4} + \frac{1}{R_5} + \frac{1}{R_3}\right)u_b = -\frac{v_5}{R_5}$$

On the left hand side of the equation of node a, the coefficient of u_a is the sum of the conductances of the individual resistor branches connected to node a. This is called the self-conductance of node a, denoted by G_{aa}. The coefficient of u_B is the negative value of the sum of the conductances corresponding to the resistor branches connected between node a and node b. The conductance between these nodes is defined as the mutual conductance between node a and node b and denoted by G_{ab}. If no resistor branch is connected between node a and node b, the mutual conductance is zero. Apparently, we have $G_{ab}=G_{ba}$.

The right hand side of the equation corresponding to node a is a constant, which equals the sum of the source currents entering node a and is denoted by i_{sA}. If no current source is connected to node a, we have $i_{sa}=0$.

The node voltage equation for node b bears the same structure. Therefore, from observing the configuration of the circuit, we can formulate the node voltage equations by applying the principles stated above, without the need to derive the equations from KCL, KVL and Ohm's law. The generic form of a node voltage equation is

$$G_{kk}u_k + \sum_{j \neq k} G_{kj}u_j = i_{sk} \tag{3.4}$$

It should be noted that when applying the node voltage method, the sources in the circuit should be current sources, and the voltage sources (in connection with a resistor) should be converted to the equivalent current sources, as shown in Fig. 3.1(b).

Example 3.1: Consider the circuit shown in Fig. 3.2, in which node d is set as the reference node. List the node voltage equations for the rest nodes.

Solution:

The circuit has three independent nodes, a, b, and c. We can list their node voltage equations directly from the circuit as follows

节点 a：
$$\left(\frac{1}{R_1}+\frac{1}{R_2}+\frac{1}{R_6}\right)u_a -\frac{1}{R_6}u_b -\frac{1}{R_2}u_c =-\frac{v_{s1}}{R_1}-\frac{v_{s2}}{R_2}$$

节点 b：
$$-\frac{1}{R_6}u_a +\left(\frac{1}{R_3}+\frac{1}{R_4}+\frac{1}{R_6}\right)u_b -\frac{1}{R_4}u_c =\frac{v_{s3}}{R_3}$$

节点 c：
$$-\frac{1}{R_2}u_a -\frac{1}{R_4}u_b +\left(\frac{1}{R_2}+\frac{1}{R_4}+\frac{1}{R_5}\right)u_c =\frac{v_{s2}}{R_2}+\frac{v_{s5}}{R_5}$$

【例 3.2】 列出图 3.3 所示电路中各节点电压方程。

解　如图 3.3 选择参考节点，对四个独立节点分别赋予 1、2、3、4 的名称。由观察法列写节点电压方程。

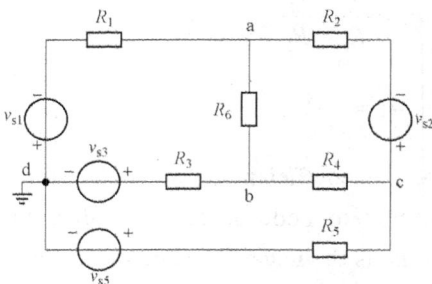

图 3.2　例 3.1 的电路图

Fig. 3.2　Circuit diagram for Example 3.1

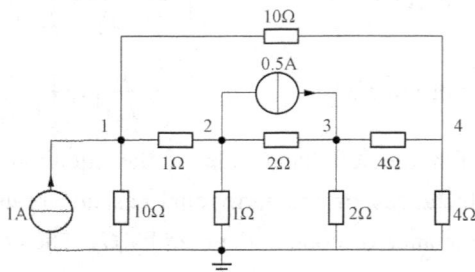

图 3.3　例 3.2 的电路图

Fig. 3.3　Circuit diagram for Example 3.2

节点 1：$\qquad (1+0.1+0.1)u_1 -u_2 -0.1u_4 =1$

节点 2：$\qquad -u_1 +(1+1+0.5)u_2 -0.5u_3 =-0.5$

节点 3：$\qquad -0.5u_2 +(0.5+0.5+0.25)u_3 -0.25u_4 =0.5$

节点 4：$\quad -0.1u_1 -0.25u_3 +(0.1+0.25+0.25)u_4 =0$

如果电路只含有两个节点，所有支路都连接在两个节点之间，如图 3.4 所示，则利用节点电压方程可给出节点电压的求解公式

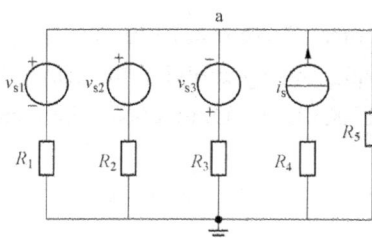

图 3.4　只含有两个节点的电路

Fig.3.4　A circuit containing only twe nodes

$$u_a =\frac{\sum i_{sa}}{G_{aa}}=\frac{\dfrac{v_{s1}}{R_1}+\dfrac{v_{s2}}{R_2}-\dfrac{v_{s3}}{R_3}+i_s}{\dfrac{1}{R_1}+\dfrac{1}{R_2}+\dfrac{1}{R_3}+\dfrac{1}{R_5}} \qquad (3.5)$$

在电路理论中，式（3.5）为弥尔曼定理。

思考题 3.1： 为何式（3.5）的分母中没有 $\dfrac{1}{R_4}$？

3.2　网孔电流法

电路分析的目的是确定电路中各条支路的电压和电流。网孔电流法是先假想电路中有沿各网孔连续流动的电流即网孔电流，然后根据 KVL 对各网孔列出关于网孔电流的线性方程组，解出各网孔电流，最后根据 KCL 解出各支路电流。这种方法也是求解复杂平面电路最基本、最直接的方法。

For node a:
$$\left(\frac{1}{R_1}+\frac{1}{R_2}+\frac{1}{R_6}\right)u_a-\frac{1}{R_6}u_b-\frac{1}{R_2}u_c=-\frac{v_{s1}}{R_1}-\frac{v_{s2}}{R_2}$$

For node b:
$$-\frac{1}{R_6}u_a+\left(\frac{1}{R_3}+\frac{1}{R_4}+\frac{1}{R_6}\right)u_b-\frac{1}{R_4}u_c=\frac{v_{s3}}{R_3}$$

For node c:
$$-\frac{1}{R_2}u_a-\frac{1}{R_4}u_b+\left(\frac{1}{R_2}+\frac{1}{R_4}+\frac{1}{R_5}\right)u_c=\frac{v_{s2}}{R_2}+\frac{v_{s5}}{R_5}$$

Example 3.2: List the node voltage equations for circuit shown in Fig. 3.3

Solution:

Select the bottom node as the reference node. For the four independent nodes 1~4, we can write the following node voltage equations based on the configuration of the circuit.

For node 1:　　　　　　　$(1+0.1+0.1)u_1-u_2-0.1u_4=1$

For node 2:　　　　　　　$-u_1+(1+1+0.5)u_2-0.5u_3=-0.5$

For node 3:　　　　　　　$-0.5u_2+(0.5+0.5+0.25)u_3-0.25u_4=0.5$

For node 4:　　　　　　　$-0.1u_1-0.25u_3+(0.1+0.25+0.25)u_4=0$

Suppose a circuit contains only two nodes and all the branches are connected between them, as shown in Fig. 3.4. In this case, the voltage of node A can be solved by using the node voltage equation

$$u_a=\frac{\sum i_{sa}}{G_{aa}}=\frac{\dfrac{v_{s1}}{R_1}+\dfrac{v_{s2}}{R_2}-\dfrac{v_{s3}}{R_3}+i_s}{\dfrac{1}{R_1}+\dfrac{1}{R_2}+\dfrac{1}{R_3}+\dfrac{1}{R_5}}\tag{3.5}$$

In circuit theory, the above formula is called the Meermann theorem.

🖝 **Exercise 3.1**: Why $\dfrac{1}{R_4}$ does not appear in the denominator of (3.5)?

3.2　Mesh Current Method

Bear in mind that circuit analysis aims to find out voltages and currents of interest. Mesh current method is a useful way to solve for currents. The method first assumes there are currents flowing through the meshes, which are called mesh currents. Next, a number of equations involving mesh currents are listed according to KVL. Finally, after the mesh currents are calculated, branch currents are solved according to KCL. The mesh current method is considered as the most straight-forward way of solving complex planar circuits. We take the circuit shown in Fig. 3.5 as an example to illustrate how to apply the mesh current method.

现以图 3.5 所示电路为例，具体说明网孔电流法的分析步骤。

（1）指定各网孔电流及方向；

（2）根据 KVL，列出各网孔的回路电压方程式。

网孔 1（a-b-d-a）：$R_6 i_{m1} + R_3 i_{m1} + R_1 i_{m1} - R_6 i_{m2} - R_3 i_{m3} = -v_{s1} - v_{s3}$

网孔 2（a-c-b-a）：$R_2 i_{m2} + R_4 i_{m2} + R_6 i_{m2} - R_4 i_{m3} - R_6 i_{m1} = v_{s2}$

网孔 3（b-c-d-b）：$R_4 i_{m3} + R_5 i_{m3} + R_3 i_{m3} - R_4 i_{m2} - R_3 i_{m1} = v_{s3} - v_{s5}$

（3）联立求解上述的独立方程，便可得到各网孔电流；

（4）根据 KCL 求出各支路电流：

$$i_1 = -i_{m1}$$

$$i_2 = i_{m2}$$

$$i_3 = i_{m3} - i_{m1}$$

$$i_4 = i_{m3} - i_{m2}$$

$$i_5 = -i_{m3}$$

$$i_6 = i_{m1} - i_{m2}$$

（5）校验计算结果是否正确。一般可以用功率平衡（能量守恒）关系校验。

【例 3.3】 电路如图 3.6 所示，已知 $R_1 = 7\,\Omega$，$R_2 = 11\,\Omega$，$R_3 = 7\,\Omega$，$v_{s1} = 70\,\text{V}$，$v_{s2} = 6\,\text{V}$，求支路电流 i_1、i_2、i_3。

解 （1）指定网孔电流及方向，如图 3.6 所示。

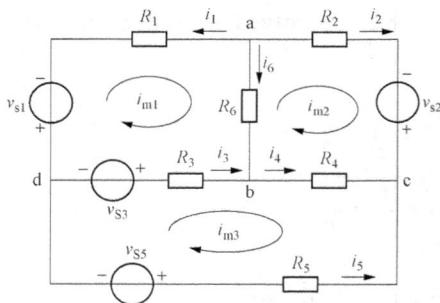

图 3.5 网孔电流法举例

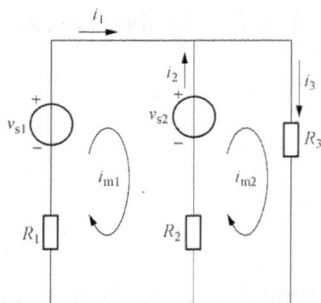

Fig. 3.5 An illustration of applying the mesh current method

图 3.6 例 3.3 的电路图

Fig. 3.6 Circuit diagram for Example 3.3

（2）根据 KVL 列出网孔电流方程式

$$R_1 i_{m1} + R_2 i_{m1} - R_2 i_{m2} = v_{s1} - v_{s2}$$

$$R_2 i_{m2} + R_3 i_{m2} - R_2 i_{m1} = v_{s2}$$

（3）代入电路参数，联立求解以上方程，得网孔电流

$$i_{m1} = 6\,\text{A}$$

$$i_{m2} = 4\,\text{A}$$

（4）根据 KCL 得各支路电流

$$i_1 = i_{m1} = 6\,\text{A}$$

$$i_3 = i_{m2} = 4\,\text{A}$$

$$i_2 = i_{m2} - i_{m1} = -2\,\text{A}$$

(1) Assign the mesh currents and their reference directions;

(2) For each mesh, list the KVL equations that involve mesh currents:

For mesh 1 (a-b-d-a): $R_6 i_{m1} + R_3 i_{m1} + R_1 i_{m1} - R_6 i_{m2} - R_3 i_{m3} = -v_{s1} - v_{s3}$

For mesh 2 (a-c-b-a): $R_2 i_{m2} + R_4 i_{m2} + R_6 i_{m2} - R_4 i_{m3} - R_6 i_{m1} = v_{s2}$

For mesh 3 (b-c-d-b): $R_4 i_{m3} + R_5 i_{m3} + R_3 i_{m3} - R_4 i_{m2} - R_3 i_{m1} = v_{s3} - v_{s5}$

(3) Solve for the above equation system to obtain the mesh currents;

(4) Calculate the branch currents based on KCL;

$$i_1 = -i_{m1}$$

$$i_2 = i_{m2}$$

$$i_3 = i_{m3} - i_{m1}$$

$$i_4 = i_{m3} - i_{m2}$$

$$i_5 = -i_{m3}$$

$$i_6 = i_{m1} - i_{m2}$$

(5) Double check the results, often by applying the law of conservation of energy (the total power supplied equals to the total power absorbed).

Example 3.3: For the circuit shown in Fig. 3.6, where $R_1 = 7\,\Omega$, $R_2 = 11\,\Omega$, $R_3 = 7\,\Omega$, $v_{s1} = 70$ V, and $v_{s2} = 6$ V, solve for currents i_1, i_2, i_3.

Solution:

(1) Assign the mesh currents and their reference directions as labeled in Fig. 3.6;

(2) For each mesh, list the corresponding KVL equations

$$R_1 i_{m1} + R_2 i_{m1} - R_2 i_{m2} = v_{s1} - v_{s2}$$

$$R_2 i_{m2} + R_3 i_{m2} - R_2 i_{m1} = v_{s2}$$

(3) Substitute the parameters of the resistances and the voltage sources into the above equations and solve them simultaneously to obtain the mesh currents

$$i_{m1} = 6 \text{ A}$$

$$i_{m2} = 4 \text{ A}$$

(4) Solve for the branch currents by applying KCL

$$i_1 = i_{m1} = 6 \text{ A}$$

$$i_3 = i_{m2} = 4 \text{ A}$$

$$i_2 = i_{m2} - i_{m1} = -2 \text{ A}$$

(5) Check if the results are correct by applying the law of conservation of energy. The algebraic sum of the powers is

（5）校验计算结果是否正确。电路的总功率的代数和为

$$R_1 i_1^2 + R_2 i_2^2 + R_3 i_3^2 - i_1 v_{s1} - i_2 v_{s2} = 7 \times 6^2 + 11 \times (-2)^2 + 7 \times 4^2 - 6 \times 70 - (-2) \times 6 = 0$$

满足功率平衡条件。

3.3　叠加原理

叠加原理是线性电路普遍适用的基本定理。所谓线性电路是指仅包含独立电源、线性受控源和线性电阻的电路。

当电路中存在多个独立（理想）电源时，电路中所有的电压、电流都是由这些独立电源共同作用所产生的。对于线性电路，所产生的电压、电流与电源满足线性关系（可加性和齐次性），利用线性关系可将多电源共同作用的电路问题，分解为多个单电源单独作用的电路问题，这就是叠加原理。

> 包含多个独立电源的线性电路，要确定电路中任意的电压或电流，可分别计算每个独立电源单独作用时（其他独立电源置零，将电压源置为短路，将电流源置为开路）产生的电压或电流，各电源共同作用产生的总电压或总电流为各独立电源单独作用时产生的分电压或分电流的代数和。

应用叠加原理可以把一个复杂电路分解成几个简单电路来研究（见图 3.7），然后将这些简单电路的计算结果叠加，便可求得原来电路中的电流或电压

$$i_1 = i_1' + i_1'' \qquad i_2 = i_2' + i_2'' \qquad i_3 = i_3' + i_3''$$

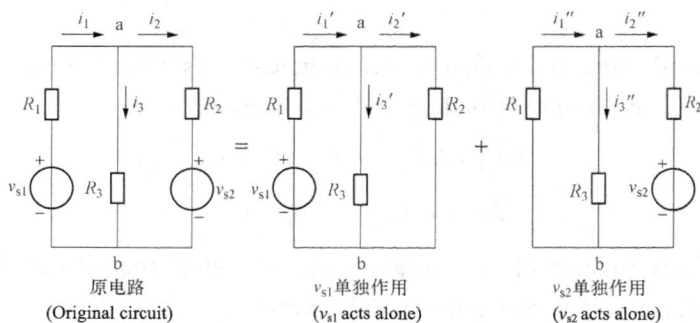

图 3.7　叠加原理示意图

Fig. 3.7　An illustration to the superposition principle

【例 3.4】对如图 3.8（a）所示电路，应用叠加原理求电压 v。

解　12 V 电压源单独作用时，3 A 电流源开路，如图 3.8（b）所示电路，此时 3 Ω 电阻上的电压为

$$v' = -\frac{3}{3+6} \times 12 = -4 \text{ V}$$

3 A 电流源单独作用时，12 V 电压源短路，如图 3.8（c）所示电路，此时 3 Ω 电阻上的电压为

$$v'' = 3 \times \frac{6 \times 3}{3+6} = 6 \text{ V}$$

$$R_1i_1^2 + R_2i_2^2 + R_3i_3^2 - i_1v_{s1} - i_2v_{s2} = 7 \times 6^2 + 11 \times (-2)^2 + 7 \times 4^2 - 6 \times 70 - (-2) \times 6 = 0$$

which suffices the law of conservation of energy.

3.3 Superposition Principle

Superposition principle is applicable to linear circuits, which are composed of independent sources, linear dependent sources, and linear resistors.

Suppose in a circuit there are two or more independent sources. The voltages and currents appear in response to the combined function of the independent sources. If it is a linear circuit, the voltages and currents further have a linear relationship (the additive property and homogeneity). In this case, a circuit driven by a number of sources can be decomposed into a number of circuits driven by a single source, respectively. This is the superposition principle.

For a linear circuit that contains more than one independent source, in order to solve for a current or voltage, we can calculate the sum of the currents or voltages in response to each of the independent sources acting individually. (If one of the independent sources acts individually, all the other independent sources are zeroed; voltage sources are replaced by short circuits and current sources become open circuits.) The algebraic sum of the currents or voltages driven by each of the independent sources acting individually is the current or voltage driven by the independent sources acting jointly.

By applying the superposition principle, we can decompose a complex circuit into a number of simple circuits, as illustrated in Fig. 3.7, and then add up the responses of the simple circuits to obtain the currents or voltages of the original circuit.

$$i_1 = i_1' + i_1'' \qquad i_2 = i_2' + i_2'' \qquad i_3 = i_3' + i_3''$$

Example 3.4: Use the superposition principle to solve for voltage v in the circuit shown in Fig. 3.8(a).

Solution:

First we analyze the circuit excited by the voltage source only. The current source is replaced by an open circuit, as shown in Fig. 3.8(b). In this case, the voltage across the 3 Ω resistance is

$$v' = -\frac{3}{3+6} \times 12 = -4 \text{ V}$$

Then we consider the circuit with only the current source activated. The voltage source becomes a short circuit, as shown in Fig. 3.8(c). Hence, the voltage across the 3 Ω resistance is

$$v'' = 3 \times \frac{6 \times 3}{3+6} = 6 \text{ V}$$

叠加，确定总电压为

$$v = v' + v'' = 2\text{ V}$$

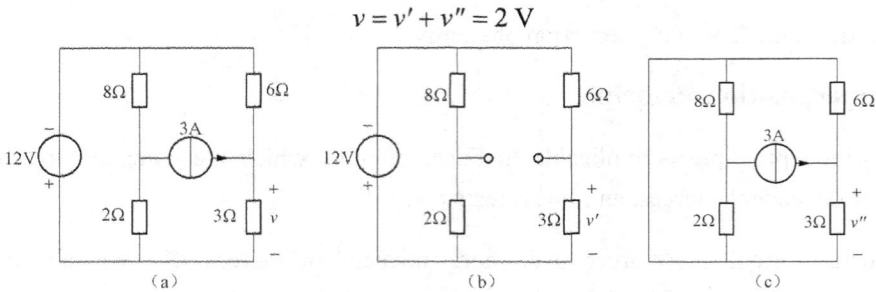

图 3.8 例 3.4 的电路图

（a）原电路；（b）仅电压源作用；（c）仅电流源作用

Fig. 3.8 Circuit diagram for Example 3.4

(a) Original circuit; (b) Voltage source acts alone; (c) Current source acts alone

应用叠加原理要注意的问题：

1）叠加原理只适用于求解线性电路的电压和电流响应。

2）叠加时只将独立电源分别考虑，电路其他部分（包括受控电源）的结构和参数不变。

3）各独立电源单独作用分析时，应保留各支路电流、电压的参考方向。确保最后叠加时各分量具有统一的参考方向。

4）叠加原理是电路线性关系的应用，由于电路中功率与电压或电流的关系为二次函数，不具有线性关系，因此，叠加原理只能用于电压或电流的计算，不能直接用来计算功率。

5）运用叠加原理求解时也可以把电源分组求解，每个分电路的电源个数可能不止一个，如图 3.9 所示，将独立电源分成电压源与电流源两组。

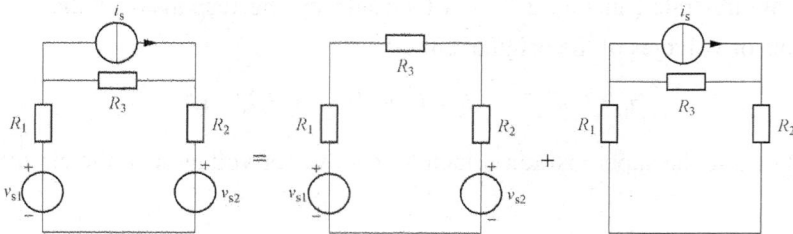

图 3.9 叠加原理中独立电源分组

Fig. 3.9 Independent sources can be grouped when applying superposition principle

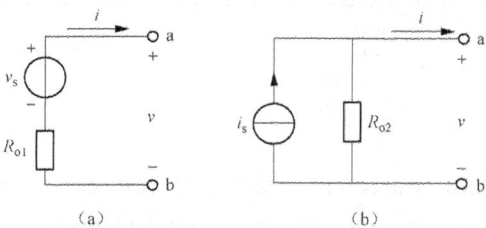

图 3.10 实际电源模型

（a）实际电压源模型；（b）实际电流源模型

Fig. 3.10 Actual Source Model

(a) Current source; (b) Voltage source

3.4 电源等效变换

实际电源都是有内阻的，其模型如图 3.10 所示。在第 1 章中所涉及电压源是图 3.10（a）中其内阻 R_{o1} 为零的理想情况，即理想电压源；第一章所涉及电流源是图 3.10（b）中其内阻 R_{o2} 为无穷大的理想情况，即理想电流源。理想电压源和理想电流源在实际中是不存在的。

Thus, the total voltage v of in Fig. 3.8(a) is

$$v = v' + v'' = 2 \text{ V}$$

Some notes on applying the superposition principle:

1. Superposition principle can only be used to solve for currents or voltages of linear circuits.

2. The circuit is decomposed according to the independent sources, and the rest elements (including controlled sources) and the configuration of the circuit remain unchanged.

3. When analyzing the decomposed circuits, the reference directions of branch currents and the reference polarities of voltages in these circuits should be the same, so that the currents and voltages have unified reference directions and polarities when summed up to obtain their total responses.

4. Superposition principle is based on the linear property of a circuit, therefore it can only be applied to solve for currents and voltages. As power is the quadratic function of current or voltage, superposition principle cannot be applied directly to calculate power.

5. The sources can be decomposed into groups when applying superposition principle and a group may contain more than one source. An illustration is shown in Fig. 3.9, where independent sources are grouped into voltage sources and current sources.

3.4   Source Transformations

Unlike ideal sources, practical sources have internal resistance. Their circuit models are shown in Fig. 3.10. The ideal voltage source we introduced in Chapter 1 has zero internal resistance, i.e. $R_{o1}=0$ in Fig. 3.10(a); the ideal current source we introduced in Chapter 1 has zero infinite resistance, i.e. $R_{o1}=\infty$ in Fig.3.10 (b). It is well known that, ideal sources do not exist in practice.

Georg Simon Ohm (16 March 1789 – 6 July 1854) is a German physicist and mathematician. As a school teacher, Ohm began his research with the new electrochemical cell, invented by Italian scientist Alessandro Volta. Using equipment of his own creation, Ohm found that there is a direct proportionality between the potential difference (voltage) applied across a conductor and the resultant electric current. This relationship is known as Ohm's law.

实际电压源与电流源只要对外电路的作用相同,即它对外电路提供的电压 v 和电流 i 相同,则它对外电路产生的效果是一样的,它们之间就可以进行变换,这就是电源等效变换的概念。

图 3.10（a）所示电路的外特性为

$$v = v_s - R_{o1}i \tag{3.6}$$

图 3.10（b）所示电路的外特性为

$$i = i_s - \frac{v}{R_{o2}} \text{或} v = R_{o2}i_s - R_{o2}i \tag{3.7}$$

要使电压源模型与电流源模型互相等效,它们的外特性必须相同,因此,两种实际电源模型的等效转换条件为

$$R_{o1} = R_{o2} = R_o \tag{3.8}$$

$$v_s = R_o i_s \tag{3.9}$$

也就是说,在满足以上两式的条件下,电压源和电流源互相变换,对外电路产生的效果是一样的。

【例 3.5】 电路如图 3.11（a）所示,已知 $R_1 = 3\,\Omega$, $R_2 = 2\,\Omega$, $R_3 = 12\,\Omega$, $i_s = 4\,\text{A}$, $v_s = 6\,\text{V}$, 求 v_{ab}。

解 由于与理想电压源 v_s 并联的元件不影响电压源的外特性,因此电阻 R_1 不影响外电路。应用电源等效变换将电流源模型（i_s 与 R_3 并联）变换为电压源模型（$R_3 i_s$ 与 R_3 串联）,如图 3.11（b）所示。列出 KVL 方程

$$-v_s + (R_2 + R_3)i + R_3 i_s = 0$$

求解得到 $i = -3\,\text{A}$。根据欧姆定律有

$$v_{ab} = R_2 i = -6\,\text{V}$$

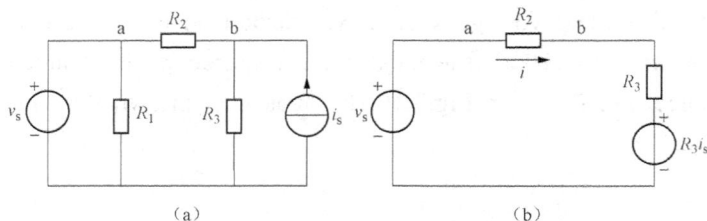

图 3.11 例 3.5 的电路图

（a）原电路；（b）将电流源等效为电压源

Fig. 3.11　Circuit diagrams for Example 3.5

(a) Original circuit; (b) Source transformation

【例 3.6】 电路如图 3.12（a）所示,已知 $R_1 = 4\,\Omega$, $R_2 = 2\,\Omega$, $R_3 = 4\,\Omega$, $R_4 = 3\,\Omega$, $R_5 = 5\,\Omega$, $i_{s1} = 4\,\text{A}$, $i_{s2} = 3\,\text{A}$, $v_s = 8\,\text{V}$, 求 R_5 支路的电流 i_5。

解 与理想电流源串联的元件不影响电流源的外特性,因此电阻 R_2 不影响电流源 i_{s1} 的外特性。利用电源等效变换将电压源（v_s 与 R_1 串联）变换为电流源模型（$\frac{v_s}{R_1}$ 与 R_1 并联）,如图 3.12（b）所示。将并联的电流源等效合并,电阻 R_1 和 R_3 并联等效,电路 3.12（b）变换为图 3.12（c）。进一步将电流源变换为电压源如图 3.12（d）所示,图中电压源 v 的电压为

$$v = \left(\frac{v_s}{R_1} + i_{s1}\right)(R_1 /\!/ R_3) = 12\,\text{V}$$

If a voltage source and a current source have identical behavior on an external circuit, i.e. they provide the same voltage v and the same current i to the external circuits, they are considered as equivalent. This is called source transformation.

The external behavior of circuit shown in Fig. 3.10(a) is

$$v = v_s - R_{o1}i \tag{3.6}$$

The external behavior of circuit shown in Fig. 3.10(b) is

$$i = i_s - \frac{v}{R_{o2}} \quad \text{or} \quad v = R_{o2}i_s - R_{o2}i \tag{3.7}$$

In order to make sure their external behavior is identical, we have

$$R_{o1} = R_{o2} = R_o \tag{3.8}$$

$$v_s = R_o i_s \tag{3.9}$$

Under this condition, the two sources are equivalent. The voltage source can be transformed into a current source, and vice versa, and their behavior to external circuits is the same.

Example 3.5: The circuit is given in Fig. 3.11(a), where $R_1 = 3$ Ω, $R_2 = 2$ Ω, $R_3 = 12$ Ω, $i_s = 4$ A, $v_s = 6$ V. Solve for v_{ab}.

Solution:

As resistance R_1 is connected in parallel to the ideal voltage source v_s, it does not influence the external behavior of v_s. Therefore, R_1 can be removed for the calculation of v_{ab}. According to source transformation principles, the practical current source (i_s is connected in parallel to R_3) can be transformed to a practical voltage source ($R_3 i_s$ is connected in series with R_3), as shown in Fig. 3.11(b). According to KVL, we have

$$-v_s + (R_2 + R_3)i + R_3 i_s = 0$$

Solve for i and we get $i = -3$A. Applying Ohm's law to R_2 we have

$$v_{ab} = R_2 i = -6 \text{ V}$$

Example 3.6: For the circuit shown in Fig. 3.12(a), where $R_1 = 4$ Ω, $R_2 = 2$ Ω, $R_3 = 4$ Ω, $R_4 = 3$ Ω, $R_5 = 5$ Ω, $i_{s1} = 4$ A, $i_{s2} = 3$ A, $v_s = 8$ V, find current i_5.

Solution:

As resistor R_2 is connected in series with the ideal current source i_{s1}, it does not influence the external behavior of i_{s1}. Therefore, R_2 can be removed from the circuit, as shown in Fig. 3.12(b). According to source transformation principles, the practical voltage source (v_s connected in series with R_1) can be transformed to a practical current source ($\frac{v_s}{R_1}$ connected inparallel with R_1), as shown in Fig. 3.12(b). Combining the parallel connected current sources and resistors, respectively, the equivalent circuit is given in Fig. 3.12(c). Next, we transform the current sources to voltage sources correspondingly and obtain the equivalent circuit as shown in Fig. 3.12(d), where the voltage of voltage source v can be calculated from

$$v = \left(\frac{v_s}{R_1} + i_{s1} \right)(R_1 /\!/ R_3) = 12 \text{ V}$$

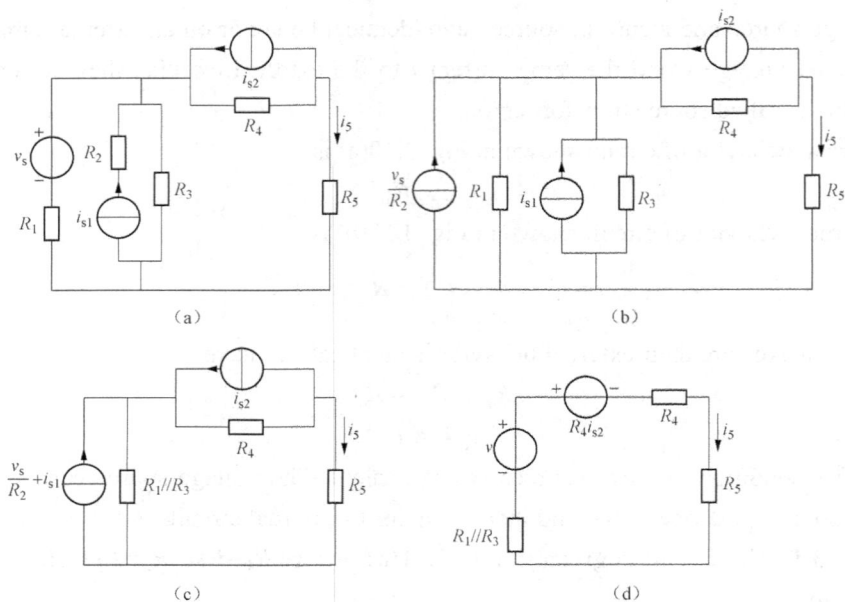

图 3.12 例 3.6 电路图

（a）原电路；（b）电压源 v_s 等效为电流源；（c）并联电流源和并联电阻的等效合并；（d）电流源等效为电压源

Fig. 3.12 Circuit diagram for Example 3.6

(a) Original circuit; (b) Voltage source v_s transformed to a current source; (c) Combination of parallel connected current sources and resistors; (d) Current sources transformed to voltage sources

由 KVL 和欧姆定律可列出如下方程

$$-v + i_5\left(R_1 /\!/ R_3\right) + i_5 R_4 + i_{s2} R_4 + i_5 R_5 = 0$$

解得 $i_5 = 0.3\,\text{A}$ 。

☞ **思考题 3-2：** 电压源和电流源在等效变换时其极性和方向是怎样对应的？

☞ **思考题 3-3：** 理想电压源和理想电流源之间能够等效变换吗？

3.5 戴维南等效电路

任何一个如图 3.13（a）所示的线性有源二端网络，对外电路而言，可等效为一个电压为 v_{oc} 的理想电压源与内阻 R_0 相串联的电压源，如图 3.13（b）所示。这就是戴维南等效电路。等效理想电压源的电压 v_{oc} 就是有源二端网络的开路电压，即将负载断开后 a、b 两端之间的电压，如图 3.13（c）所示。内阻 R_0 等于有源二端网络中所有独立电源置零（理想电压源为短路，理想电流源为开路）后所得到的无源二端网络在 a、b 两端看进去的等效电阻，如图 3.13（d）所示。该电阻也称为戴维南等效电阻。

【例 3.7】 电路如图 3.14（a）所示，已知 $R_1 = 20\,\Omega$ ，$R_2 = 30\,\Omega$ ，$R_3 = 30\,\Omega$ ，$R_4 = 20\,\Omega$ ，$v = 10\,\text{V}$ 。求当 $R_5 = 10\,\Omega$ 时 i_5 的值。

解 利用戴维南定理进行计算。将电阻 R_5 作为负载，电路其他部分作为有源二端网络。首先确定等效电源参数。有源二端网络的开路电压 v_{oc}，如图 3.14（b）所示，可由下式计算

$$v_{oc} = v\frac{R_2}{R_1 + R_2} - v\frac{R_4}{R_3 + R_4} = 2\,\text{V}$$

According to KVL and Ohm's law, we have

$$-v + i_5(R_1 \text{ // } R_3) + i_5 R_4 + i_{s2} R_4 + i_5 R_5 = 0$$

Solve the above equation we get $i_5 = 0.3 \text{ A}$.

☞ **Exercise 3.2**:   When applying source transformations, how should we determine the direction of the current source and the polarity of the voltage source?

☞ **Exercise 3.3**:   Can an ideal voltage source be transformed to an ideal current source?

3.5   Thévenin Equivalent Circuit

Any two-terminal linear network that includes sources and resistors, as shown in Fig. 3.13(a), can be replaced by the series combination of an ideal voltage source whose voltage is v_{oc} and an internal resistor whose resistance is R_0. As shown in Fig. 3.13(b), the two two-terminal networks are identical to the external circuit. This is the so-called Thévenin equivalent circuit. The voltage of the ideal voltage source, v_{oc}, is the open source voltage of the original active two-terminal network, which is the voltage across node a and node b when the load is removed from the circuit, as shown in Fig. 3.13(c). The value of the internal resistance, R_0, is the equivalent resistance between node a and node b when all the independent sources in the active two-terminal network are zeroed (ideal voltage sources are replaced by short circuits and ideal current sources are replaced by open circuits), as shown in Fig. 3.13(d). This resistance is also called Thévenin resistance.

Example 3.7: Find the value of i_5 when R_5=10 Ω for the circuit shown in Fig. 3.14(a), where R_1=20 Ω, R_2=30 Ω, R_3=30 Ω, R_4=20 Ω, v =10 V.

Solution:

We analyze this circuit using Thévenin equivalent. Resistor R_5 is considered as the load and the rest of the original circuit is considered as the active two-terminal network. First, we need to determine the value of Thévenin voltage. As shown in Fig. 3.14(b), the open circuit voltage of the active two-terminal network, v_{oc}, can be calculated from

$$v_{oc} = v \frac{R_2}{R_1 + R_2} - v \frac{R_4}{R_3 + R_4} = 2 \text{ V}$$

Léon Charles Thévenin (30 March 1857 – 21 September 1926) was a French telegraph engineer who extended Ohm's law to the analysis of complex electrical circuits. On the basis of Kirchhoff's laws and Ohm's law, he proposed the Thévenin equivalent formula, which made it possible to calculate currents in more complex electrical circuits and allowing people to reduce complex circuits into simpler circuits called Thévenin's equivalent circuits.

为确定戴维南等效电路的内阻，将理想电压源替换为短路，得到如图 3.14（c）所示的无源二端网络，则

$$R_0 = R_1 /\!/ R_2 + R_3 /\!/ R_4 = 24\ \Omega$$

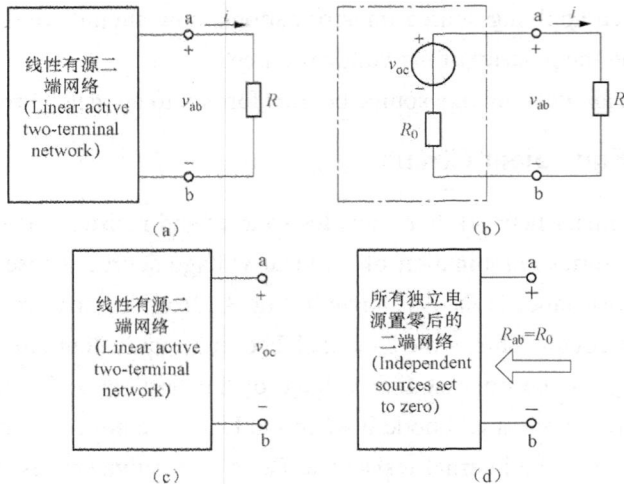

图 3.13　戴维南等效电路

Fig. 3.13　Thévenin equivalent circuit

最后，将戴维南等效电路应用到原电路得图 3.14（d）所示电路，计算 i_5：

$$i_5 = \frac{v_{oc}}{R_0 + R_5} = \frac{2}{24 + 10} = 0.059\ \text{A}$$

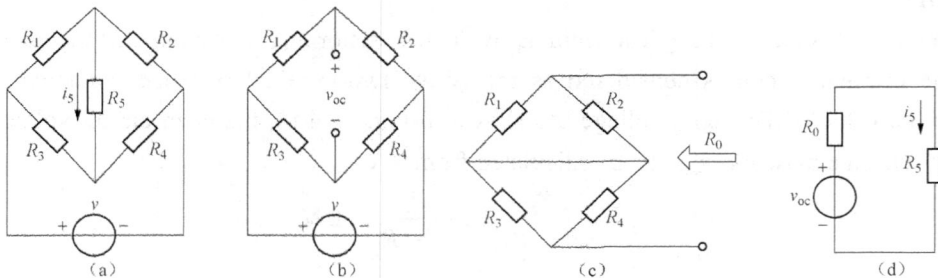

图 3.14　例 3.7 的电路图

（a）原电路；（b）求开路电压 v_{oc}；（c）求戴维南电阻 R_0；（d）戴维南等效电路

Fig. 3.14　Circuit diagram for Example 3.7

(a) Original circuit; (b) Open circuit voltage v_{oc}; (c) Thévenin resistance R_0; (d) Thévenin equivalent circuit

【例 3.8】　如图 3.15（a）所示，当开关 S 断开时，电压表读出的电压为 18 V；当 S 闭合时，电流表中读出的电流为 1.8 A。试求有源二端网络的戴维南等效电路。

解　如果不考虑电压表的内阻（视电压表内阻为无穷大），则当开关 S 断开时电压表测得的即是有源二端网络的开路电压 v_{oc}=18 V。

同样地，如果不考虑电流表内阻（视电流表内阻为零），设有源二端网络等效电路内阻为 R_0，则测量电流时的等效电路如图 3.15（b）所示，根据 KVL 有

$$-18 + 1.8R_0 + 9 \times 1.8 = 0$$

To find out the internal resistance of Thévenin equivalent circuit, we replace the voltage source v by a short circuit, as shown in Fig. 3.14(c). The equivalent resistance between node a and node b is

$$R_0 = R_1 \,//\, R_2 + R_3 \,//\, R_4 = 24\ \Omega$$

Replacing the original active two-terminal network by the Thévenin equivalent circuit, we obtain the circuit shown in Fig. 3.14(d). From this circuit we can easily find out i_5

$$i_5 = \frac{v_{oc}}{R_0 + R_5} = \frac{2}{24 + 10} = 0.059\ \text{A}$$

Example 3.8: For the circuit shown in Fig. 3.15(a), when switch S opens, the reading of the voltmeter is 18 V; when switch S is closed, the reading of the ammeter is 1.8 A. Find the Thévenin equivalent circuit of the active two-terminal network.

Solution:

If we neglect the internal resistance of the voltmeter (the internal resistance of the voltmeter is infinite), the reading of the voltmeter when the switch is open is taken as the open circuit voltage of the active two-terminal network, v_{oc}=18 V.

Similarly, we can neglect the internal resistance of the ammeter (the internal resistance of the ammeter is 0). When the switch is closed, the equivalent circuit is given in Fig. 3.15(b), where R_0 is the internal resistance of the equivalent circuit of the active two-terminal network. According to KVL, we have

$$-18 + 1.8R_0 + 9 \times 1.8 = 0$$

More facts about ammeters and voltmeters

An ammeter is an instrument used to measure the current in a circuit. In a circuit diagram, an ammeter is represented by a letter A in a circle. An ammeter is designed based on the fact that a current passing through a coil placed in the magnetic field of a permanent magnet causes the coil to move. The movement of the coil causes deflection of the pointer, which is linearly proportional to the current.

A voltmeter is an instrument used for measuring the voltage between two points in a circuit. The circuit diagram of a voltmeter is a letter V in a circle. The principle of designing a voltmeter is the same as that of an ammeter, where the deflection of the coil is proportional to the current, which in turn is proportional to the applied voltage. In order to disturb the circuit as little as possible, the instrument should draw a minimum of current to operate. This is achieved by connecting a sensitive ammeter in series with a high resistance.

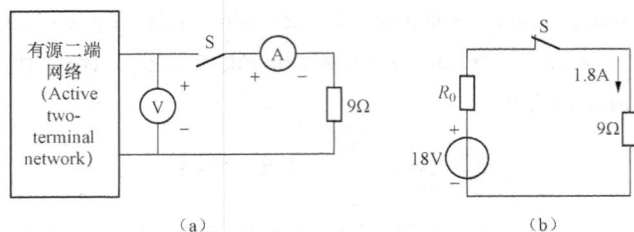

图 3.15　例 3.8 电路图

（a）原电路；（b）戴维南等效电路

Fig. 3.15　Circuit diagram for Example 3.8

(a) Original circuit; (b) Thévenin equivalent circuit

求解得到 $R_0 = 1\ \Omega$，即戴维南等效电路的内阻为 $R_0 = 1\ \Omega$。

3.6　诺顿等效电路

戴维南定理说明，任意有源二端线性网络可用一个实际的电压源模型来等效，而诺顿定理则将有源二端线性网络等效为一个实际的电流源模型。

任意有源二端线性网络，如图 3.16（a）所示，对外电路而言，等效为一个电流为 i_s 的理想电流源和内阻 R_0 相并联的电流源，如图 3.16（b）所示。等效电源中的理想电流 i_s 数值上等于该二端网络的短路电流 i_{sc}，如图 3.16（c）所示。等效电源内阻 R_0 等于有源二端网络中所有独立电源置零（理想电压源替换为短路，理想电流源替换为开路）后所得到的无源二端网络在 a、b 两端所呈现的等效电阻，如图 3.16（d）所示。

由电源等效变换原理可知，任一有源二端线性网络的戴维南等效电路和诺顿等效电路是等效的，只要求出该二端网络的开路电压 v_{oc}、短路电流 i_{sc} 和内阻 R_0 中的任意两个量，都可通过公式 $R_0 = \dfrac{v_{oc}}{i_{sc}}$ 求出第三个量，进而求得相应的戴维南或诺顿等效电路。戴维南等效电路是值为 v_{oc} 的电压源与电阻 R_0 的串联，诺顿等效电路是值为 i_{sc} 的电流源与电阻 R_0 的并联。

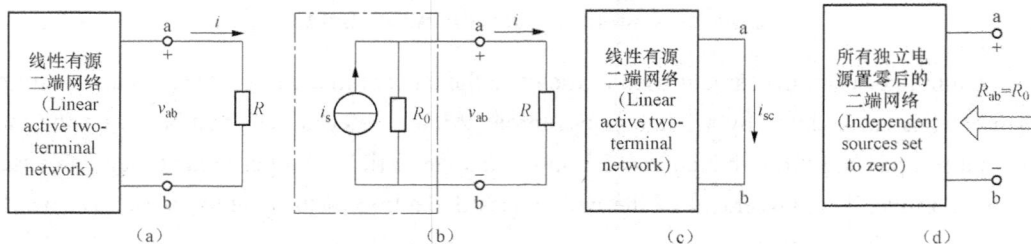

图 3.16　诺顿等效电路

Fig. 3.16　Norton equivalent circuit

【例 3.9】电路如图 3.17（a）所示，已知 $R_1 = 4\ \Omega$，$R_2 = 2\ \Omega$，$R_3 = 10\ \Omega$，$v_1 = 12\ \text{V}$，$v_2 = 24\ \text{V}$。用诺顿定理求电路中电阻 R_1 上的电流 i。

解　把原电路中除电阻 R_1 以外的部分化简为由一个理想电流源和一个内电阻并联构成的诺顿等效电路。

Solve the above equation, we get $R_0 = 1\,\Omega$, which is the Thévenin resistance.

3.6   Norton Equivalent Circuit

Thévenin equivalent circuit states that any active two-terminal linear network can be replaced by a practical voltage source model. Norton equivalent circuit, on the other hand, indicates that any active two-terminal linear network can be replaced by a practical current source model.

Any active two-terminal linear network, as shown in Fig. 3.16(a), can be replaced by the parallel combination of an ideal current source i_s and internal resistor R_0, as shown in Fig. 3.16(b). The value of the ideal current source equals to the short circuit current of the active two-terminal network, as shown in Fig. 3.16(c). The internal resistance of the equivalent current source equals to the equivalent resistance between node a and node b, when all the independent sources of the active two-terminal network are zeroed (ideal voltage sources are replaced by short circuits and ideal current sources are replaced by open circuits), as shown in Fig. 3.16(d).

According to source transformations, the Thévenin equivalent circuit of an active linear two-terminal network is equivalent to its Norton equivalent circuit. Therefore, as long as we can solve for two of the three variables—open circuit voltage v_{oc}, short circuit current i_{sc}, and internal resistance R_0, the remaining value can be calculated from $R_0 = \dfrac{v_{oc}}{i_{sc}}$. Afterwards, Thévenin equivalent circuit is built by the series connection of a voltage source of v_{oc} and a resistor of R_0, and Norton equivalent circuit is the parallel connection of a current source of i_{sc} and a resistor of R_0.

Example 3.9: Consider the circuit shown in Fig. 3.17(a), where $R_1 = 4\,\Omega$, $R_2 = 2\,\Omega$, $R_3 = 10\,\Omega$, $v_1 = 12\text{ V}$, $v_2 = 24\text{ V}$. Find the current i of resistor R_1 by applying Norton equivalent circuit.

Solution:

We can use Norton equivalent circuit, which is an ideal current source connected in parallel to an internal resistor, to replace the two-terminal network between node a and node b.

Edward Lawry Norton (28 July 1898 - 28 January 1983) was an accomplished Bell Laboratories engineer and scientist famous for developing the concept of the Norton equivalent circuit. Primarily Norton was interested in a communications circuit theory and the transmission of data at high speeds over telephone lines, yet he is best remembered for development of the dual of Thévenin's equivalent circuit, currently referred to as Norton's equivalent circuit. In 1926, he proposed the equivalent circuit using a current source and parallel resistor to assist in the design of recording instrumentation that was primarily current driven.

第一步，求等效电源的电流 i_{sc}。将 a、b 两点短路，求其短路电流 i_{sc}，如图 3.17（b）所示。注意，图 3.17（b）所示电路中，a、b 两点短路后，理想电压源 v_1 与电阻 R_2 并联，但在分析电路时却不能将 R_2 去掉，因为前面讲与理想电压源并联的支路可以等效为一个理想电压源是对其外电路而言的，而这里 R_2 是内电路的一部分。

根据电源等效变换可将图 3.17（b）中的实际电压源（v_2 与 R_3 串联）转换为实际电流源（$\dfrac{v_2}{R_3}$ 与 R_3 并联），如图 3.17（c）所示。将 R_2 和 R_3 并联后，再利用电源等效变换将电路变换为图 3.17（d）所示的形式，可直接计算短路电流为

$$i_{sc} = \frac{v_1 + \dfrac{v_2}{R_3}(R_2 /\!/ R_3)}{R_2 /\!/ R_3} = \frac{12 + 4}{1.67} = 9.58\,\text{A}$$

第二步，将有源二端网络中的所有理想电源置零，得到如图 3.17（e）所示的无源二端网络，则有

$$R_0 = R_2 /\!/ R_3 = \frac{2 \times 10}{2 + 10} = 1.67\,\Omega$$

第三步，在原电路中二端网络用其诺顿等效电路替换，如图 3.17（f）所示。通过该电路可轻松求解出电流 i

$$i = -\frac{R_0}{R_1 + R_0} i_{sc} = -\frac{1.67}{4 + 1.67} \times 9.58 = -2.82\,\text{A}$$

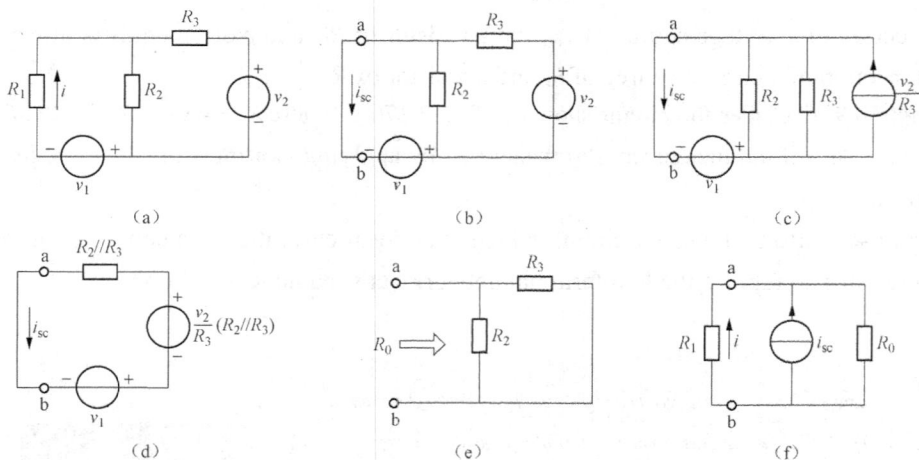

图 3.17　例 3.9 电路图

（a）原电路；（b）求短路电流 i_{sc}；（c）电源等效变换；

（d）并联电阻和电源等效变换；（e）求戴维南电阻 R_0；（f）原电路的诺顿等效电路

Fig. 3.17　Circuit diagram for Example 3.9

(a) Original circuit; (b) Short circuit current i_{sc}; (c) Source transformation; (d) Parallel resistors and source trans formation;

(e) Thévenin resistor R_0; (f) Norton equivalent circuit of the original circuit

✎ **讨论：** 含受控源电路的等效电路

由于受控源的输出受到另一支路电压或电流的控制，在分析含有受控源的线性电路时，不能像独立电源那样处理受控源，在分析过程中要注意以下几点：

First, find out the current of the equivalent current source i_{sc}. This is achieved by connecting node a and node b by a short circuit and solve for the short circuit current i_{sc}, as shown in Fig. 3.17(b). It should be noted that in this circuit, although ideal voltage source v_1 and resistor R_2 are connected in parallel, R_2 cannot be removed for circuit analysis, as we did in Example 3.5. In this case, R_2 is actually part of the internal circuit.

Based on source transformation, the practical voltage source in Fig. 3.17(b) (v_2 and R_3 connected in series) can be transformed to a practical current source ($\dfrac{v_2}{R_3}$ and R_3 connected in parallel), as shown in Fig. 3.17(c). As R_2 and R_3 are connected in parallel, the circuit can be further simplified to the one shown in Fig. 3.17(d), from which the short circuit current can be calculated

$$i_{sc} = \frac{v_1 + \dfrac{v_2}{R_3}(R_2 // R_3)}{R_2 // R_3} = \frac{12+4}{1.67} = 9.58 \text{ A}$$

Next, all the independent sources in the active two-terminal network are zeroed, so we have a network shown in Fig. 3.17(e), from which the internal resistance can be calculated

$$R_0 = R_2 // R_3 = \frac{2 \times 10}{2 + 10} = 1.67 \ \Omega$$

Finally, replace the two-terminal network in the original circuit by its Norton equivalent circuit, as shown in Fig. 3.17(f). From this circuit, it is easy for us to solve for current i

$$i = -\frac{R_0}{R_1 + R_0} i_{sc} = -\frac{1.67}{4 + 1.67} \times 9.58 = -2.82 \text{ A}$$

Discussion: Equivalent circuit of a circuit with dependent sources

The output of a dependent source is controlled by some other current or voltage. Therefore, for the analysis of a linear circuit with dependent sources, we cannot zero them as we did for independent sources. Here are some principles to follow:

（1）在对含受控源电路进行等效变换时，变换过程中不能使控制量消失，否则会留下一个没有控制量的受控源电路，使电路无法求解。

（2）在求解含受控源二端网络的等效电源电路时，注意：

1）由于只有在独立电源产生控制作用后，受控源才能够表现出电源性质，因此，如果二端网络内除了受控源外没有其他独立电源，则此二端网络的开路端电压必为 0。

2）求二端网络等效电阻时，只将网络中的独立电源置零，受控源必须保留。因此，一般不能通过电阻串并联等效来计算等效电阻。

以下通过几个例子来说明。

【例 3.10】 电路如图 3.18（a）所示，已知 $R_1 = 3\,\Omega$，$R_2 = 1\,\Omega$，若 $i_1 = 2\,\text{A}$，则 i_s 的值是多少？

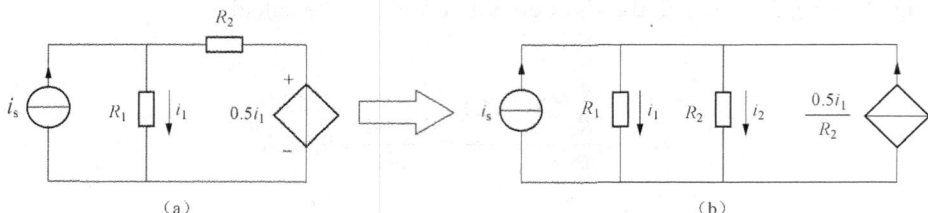

图 3.18　例 3.10 电路图

Fig. 3.18　Circuit diagram for Example 3.10

解　电路进行等效变换后如图 3.18（b）所示，由于电阻 R_1 中电流为受控电源的控制量，因此，不能再进一步将电阻作并联等效。根据并联电阻两端电压相同和欧姆定律，可以求出 i_2

$$i_2 = \frac{R_1 i_1}{R_2} = \frac{3 \times 2}{1} = 6\,\text{A}$$

再由 KCL 可求得

$$i_s = i_1 + i_2 - \frac{0.5 i_1}{R_2} = 2 + 6 - 1 = 7\,\text{A}$$

【例 3.11】 电路如图 3.19（a）所示，已知 $R_1 = 6\,\Omega$，$R_2 = 1\,\Omega$，$R_3 = 4\,\Omega$，$R_4 = 2\,\Omega$，$v_s = 9\,\text{V}$，用戴维南定理求 i_1。

解　将 i_1 所在支路以外的电路部分视为一个二端网络，用其戴维南等效电路替换。

首先将 i_1 所在支路断开，如图 3.19（b）所示，求开路电压 v_{oc}。由于此二端网络中不含有独立电源，因此，可以判断开路电压是零。事实上，在该电路中，由于 $i_1 = 0$，所以受控电流源的电流 $0.5 i_1$ 也为零，因此有 $v_{oc} = 0$。

接下来，求戴维南等效电阻。在端口 ab 间加一电流源 i_x，如图 3.19（c）所示，求出端口电压 v_x 关于 i_x 的表达式，进而求出戴维南等效电阻 $R_0 = \dfrac{v_x}{i_x}$。注意这里无需也无法求出 v_x 和 i_x 的值。根据 KCL 和欧姆定律，可以列出如下方程：

$$\begin{cases} i_2 = i_x + i_3 \\ i_3 = 0.5 i_x + i_4 \\ v_x = R_2 i_2 \\ -v_x = R_4 i_4 + R_3 i_3 \end{cases}$$

1. When performing equivalent transformations for a circuit with dependent sources, the control variables should always appear in the circuit. Otherwise, there will be a dependent source without a control variable, which makes it impossible to solve the circuit.

2. For the analysis of a two-terminal network with dependent sources.

(1) A dependent source only behaves as source when it is controlled by a voltage or current driven by independent sources. Therefore, if no independent source exists in the two-terminal network, the open circuit voltage must be 0.

(2) To find out the equivalent resistance of the two-terminal network, the independent sources should be zeroed, while the dependent sources are unchanged. Due to the existence of the dependent sources, we cannot calculate the equivalent resistance through series and parallel combinations.

We present some examples to illustrate the Thévenin and Norton equivalents of a circuit with dependent sources.

Example 3.10: For the circuit shown in Fig. 3.18(a), where $R_1 = 3\ \Omega$ and $R_2 = 1\ \Omega$. Solve for i_s when $i_1 = 2\ \text{A}$.

Solution:

After source transformation, the equivalent circuit is given in Fig. 3.18(b). As the dependent current source is controlled by i_1, we should not combine the two resistors connected in parallel. Based on the principle that parallel connected resistors share the same voltage and Ohm's law, we can solve for i_2

$$i_2 = \frac{R_1 i_1}{R_2} = \frac{3 \times 2}{1} = 6\text{A}$$

According to KCL, we can find out i_s

$$i_s = i_1 + i_2 - \frac{0.5 i_1}{R_2} = 2 + 6 - 1 = 7\text{A}$$

Example 3.11: For the circuit shown in Fig. 3.19(a), where $R_1 = 6\ \Omega$, $R_2 = 1\ \Omega$, $R_3 = 4\ \Omega$, $R_4 = 2\ \Omega$, $v_s = 9\ \text{V}$. Solve for i_1 using Thévenin equivalent.

Solution:

We use the Thévenin equivalent circuit to replace the original circuit excluding branch i_1, which is a two-terminal network.

First, we use an open circuit to replace branch i_1, as shown in Fig. 3.19(b), and find out the open circuit voltage v_{oc}. As the two-terminal network does not contain independent sources, the open circuit voltage is zero. In fact, in this network as $i_1 = 0$, the current of the dependent current source is also zero. Therefore, $v_{oc} = 0$.

Next, we find out the Thévenin resistance. We apply an independent current source i_x between node a and node b, as shown in Fig. 3.19(c), list the equation system involving v_x and i_x, and calculate the Thévenin resistance $R_0 = \dfrac{v_x}{i_x}$. Note that we need not and cannot find out the values of v_x and i_x. According to KCL and Ohm's law, we can list the following equations

$$\begin{cases} i_2 = i_x + i_3 \\ i_3 = 0.5 i_x + i_4 \\ v_x = R_2 i_2 \\ -v_x = R_4 i_4 + R_3 i_3 \end{cases}$$

联立以上方程可以解得 $R_0 = \dfrac{v_x}{i_x} = 1\ \Omega$ 。即原二端网络的戴维南等效电路为一个阻值为 1 Ω 的电阻。

将戴维南等效电路应用到原电路，如图 3.19（d）所示，可以解得

$$i_1 = -\frac{v_s}{R_1 + R_0} = -\frac{9}{6+1} = -1.3\ \mathrm{A}$$

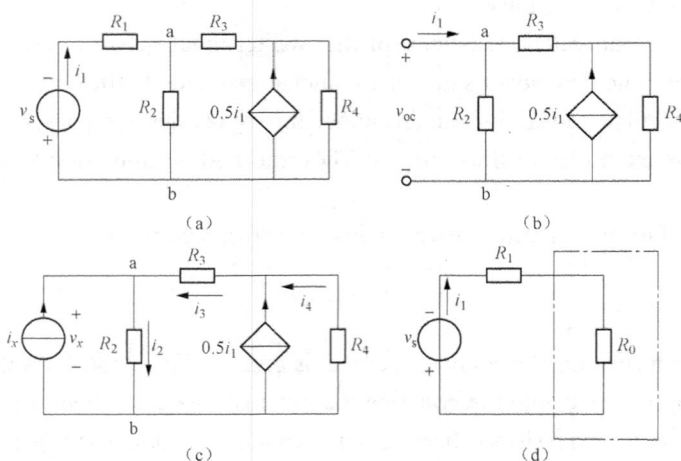

图 3.19　例 3.11 的电路图

（a）原电路；（b）求开路电压 v_{oc}；（c）求戴维南等效电阻 R_0；（d）原电路的戴维南等效电路

Fig. 3.19　Circuit diagrams for Example 3.11:

(a) Original circuit; (b) Open circuit voltage v_{oc}; (c) Thévenin resistance R_0; (d) Thévenin equivalent circuit of the original circuit

【例 3.12】　电路如图 3.20（a）所示，已知 $R_1 = 6\ \Omega$ ， $R_2 = 4\ \Omega$ ， $v_s = 20\ \mathrm{V}$ ， $i_s = 10\ \mathrm{A}$ ，求 i_2 和 v。

解　首先将含 i_2 支路断开，如图 3.20（b）所示，求端口 ab 间的开路电压 v_{oc}。根据 KVL 和 KCL 可分别列出如下方程：

$$\begin{cases} -v_s + R_1 i_1 + v_{oc} = 0 \\ i_1 = -i_s \end{cases}$$

联立求解可得 $v_{oc} = 80\ \mathrm{V}$。

接下来，将 a、b 两节点短接，如图 3.20（c）所示，求短路电流 i_{sc}。根据 KCL 和欧姆定律有

$$i_{sc} = i_1 + i_s = \frac{v_s}{R_1} + i_s = \frac{20}{6} + 10 = \frac{40}{3}\ \mathrm{A}$$

从而可计算戴维南电阻为 $R_0 = \dfrac{v_{oc}}{i_{sc}} = 6\ \Omega$ 。因此，原二端网络的戴维南等效电路为一个 80V 电压源与一个 6 Ω 电阻的串联。

将戴维南等效电路代入原电路，如图 3.20（d）所示，可求得电流 i_2 为

$$i_2 = \frac{v_{oc}}{R_0 + R_2} = \frac{80}{6+4} = 8\ \mathrm{A}$$

Solving the above equations simultaneously, we get $R_0 = \dfrac{v_x}{i_x} = 1\,\Omega$. Therefore, the Thévenin equivalent circuit of the original two-terminal network is a resistor whose resistance is $1\,\Omega$.

Substitute the original two-terminal network by its Thévenin equivalent, we obtain the circuit, as shown in Fig. 3.19(d), from which i_1 can be calculated

$$i_1 = -\frac{v_s}{R_1 + R_0} = -\frac{9}{6+1} = -1.3\ \text{A}$$

Example 3.12: For the circuit shown in Fig. 3.20 (a), where $R_1 = 6\,\Omega$, $R_2 = 4\,\Omega$, $v_s = 20\ \text{V}$, $i_s = 10\ \text{A}$. Solve for i_2 and v.

Solution:

First, we replace branch i_2 by an open circuit, as shown in Fig. 3.20(b), and solve for the open circuit voltage v_{oc}. According to KVL and KCL, we can list the following equations

$$\begin{cases} -v_s + R_1 i_1 + v_{oc} = 0 \\ i_1 = -i_s \end{cases}$$

Solving the above equations simultaneously, we get $v_{oc} = 80\ \text{V}$.

Next, we short nodes a and b, as shown in Fig. 3.20(c), to solve for the short circuit current i_{sc}. According to KCL and Ohm's law, we have

$$i_{sc} = i_1 + i_s = \frac{v_s}{R_1} + i_s = \frac{20}{6} + 10 = \frac{40}{3}\ \text{A}$$

Consequently, the Thévenin resistance can be calculated as $R_0 = \dfrac{v_{oc}}{i_{sc}} = 6\,\Omega$. The original two-terminal network is equivalent to an 80V voltage source connected in series with a $6\,\Omega$ resistor.

Substituting the Thévenin equivalent circuit into the original circuit, as shown in Fig. 3.20(d), we can solve for current i_2

$$i_2 = \frac{v_{oc}}{R_0 + R_2} = \frac{80}{6+4} = 8\ \text{A}$$

Applying KVL, KCL and Ohm's law on the original circuit, we have the following equations

$$\begin{cases} 10 i_1 + v - R_2 i_2 = 0 \\ i_1 + i_s = i_2 \end{cases}$$

Solving the above equations simultaneously, we can find out the voltage across the current source

$$v = 52\ \text{V}$$

在原电路上应用 KVL、KCL 和欧姆定律可以列出如下方程

$$\begin{cases} 10i_1 + v - R_2 i_2 = 0 \\ i_1 + i_s = i_2 \end{cases}$$

联立求解上述方程可得到电流源两端电压

$$v = 52 \text{ V}$$

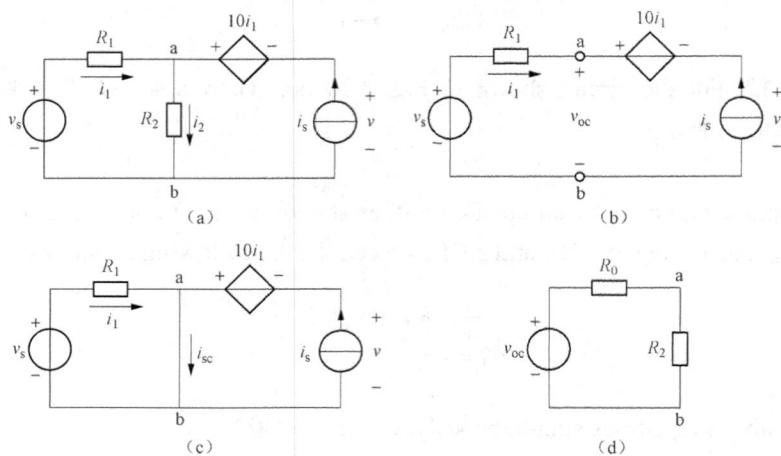

图 3.20　例 3.12 的电路图

（a）原电路；（b）求开路电压 v_{oc}；（c）求短路电流 i_{sc}；（d）原电路的戴维南等效电路

Fig. 3.20　Circuit diagrams for Example3.12

(a) Original circuit; (b) Open circuit voltage v_{oc}; (c) Short circuit current i_{sc}; (d) Thévenin equivalent circuit of the original circuit

▶ Problems

P3.1   For the circuit shown in Fig. P3.1, list the node voltage equations for nodes 1, 2 and 3, respectively.

P3.2   Circuit diagram is given in Fig. P3.2. Use node voltage method to solve for the voltage v_L across resistance R_L.

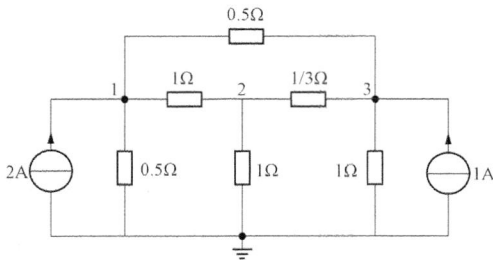

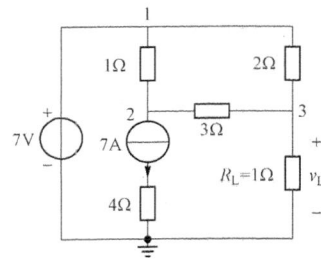

Fig. P3.1   Fig. P3.2

P3.3   For the circuit shown in Fig. P3.3, list the mesh current equations to solve for current i_5.

P3.4   For the circuit shown in Fig. P3.4, list the mesh current equations to solve for current i.

P3.5   Circuit diagram is shown in Fig. P3.5, where $v_s = 4\,\text{V}$, $i_{s1} = 4\,\text{A}$, $i_{s2} = 2\,\text{A}$, $R_1 = R_2 = 4\,\Omega$, $R_3 = R_4 = 2\,\Omega$. Use the superposition principle to solve for i_4 and i.

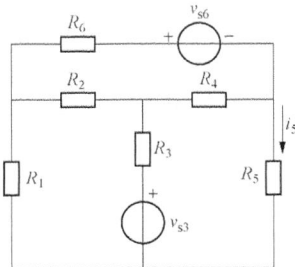

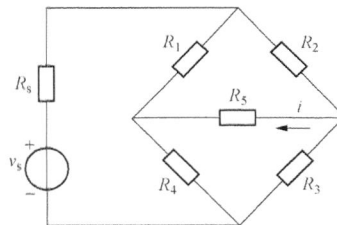

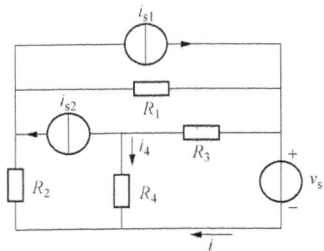

Fig. P3.3   Fig. P3.4   Fig. P3.5

P3.6   Find out the current i in Fig. P3.6 using source transformations.

P3.7   Find out the current i in Fig. P3.7 using Thévenin equivalent.

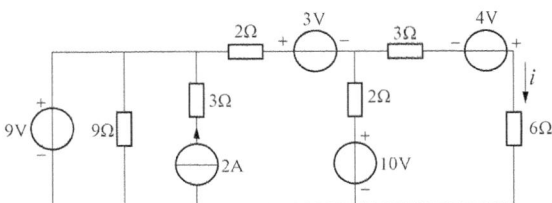

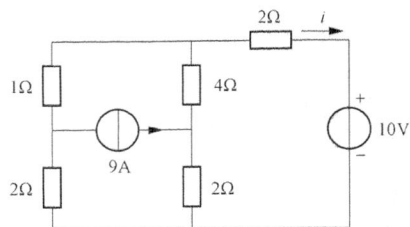

Fig. P3.6   Fig. P3.7

P3.8   For the circuit shown in Fig. P3.8, use Thévenin equivalent circuit to solve for the current i when R_x=1.2 Ω and R_x=5.2 Ω, respectively.

P3.9   For the circuit shown in Fig. P3.9, use Thévenin equivalent circuit to solve for the power consumed by the load resistance R_L.

P3.10   For the circuit shown in Fig. P3.10, use Norton equivalent circuit to solve for the voltage v across the 1 A current source.

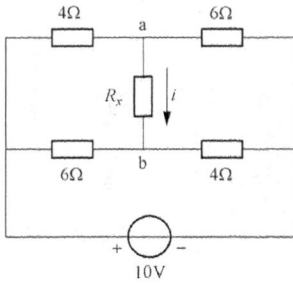

Fig. P3.8

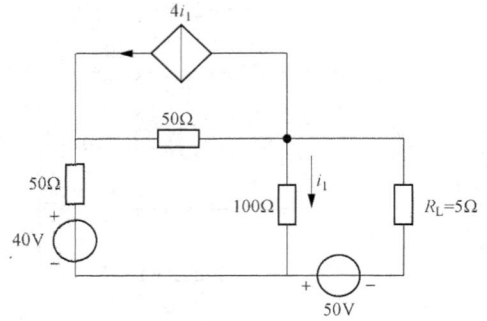

Fig. P3.9

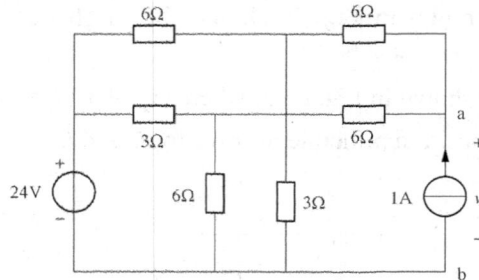

Fig. P3.10

直流稳态与暂态分析
DC Steady State
and Transient Analysis

前面几章中涉及的电路仅包括三类元件：导线、电源与电阻。在本章中，我们将学习两种新的电路元件：电容与电感。与电阻直接消耗能量不同，电容和电感是能量存储元件，它们能够存储能量，然后再将能量返回给电路。由于这两种元件的存在，直流电源的断开或接入（或信号的突然接入，通常由开关通断引起）将引起电压、电流随时间变化，这一现象称为暂态。相应地，电路结构或元件参数没有变化的状态称为稳态。可以说暂态是电路由一个稳态过渡至另一个稳态的过程。虽然本章所分析的电路包括了电容与电感这两种新元件，电路的工作状态也有稳态与暂态之分，但电路分析原则不变，在前几章中所学的分析方法与原理等均适用。

4.1　电容与电感

4.1.1　电容

电容元件是一种表征电路元件存储电荷特性的理想电路元件。实际中的电容元件结构千变万化，其中平行板电容器是一种简单常用的电容器。它通常由以一薄层绝缘材料分隔的两块平行的导体（通常是金属）极板构成，这层绝缘材料称为电介质，可以是空气、聚酯薄膜、云母、电解质等。当电流流入一端极板时，如图 4.1 所示，电流向下流入上方极板，意味着电子向上移动，并最终聚集在下方极板，使得下方极板带有负电荷，从而在电介质中建立电场。

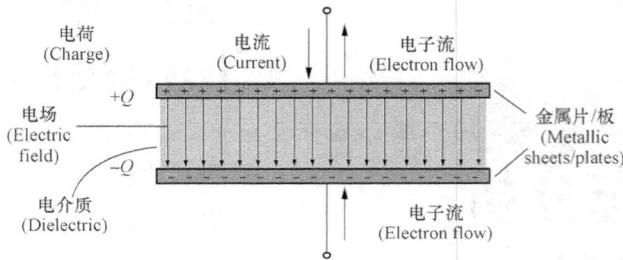

图 4.1　电容器示意图

Fig. 4.1　An illustration of a capacitor

当用电容元件描述这一现象时，电容元件所存储的电荷 q 与极板两端电压 v 呈正比

$$q = Cv \tag{4.1}$$

其中比例系数 C 即为电容元件的电容。在后续行文过程中，将不再特意区分电容元件与电容。电容的单位为法拉（简称法，F）。法拉是一个很大的单位，在大多数情况下我们所用到的电容单位都是皮法（$1\text{pF}=10^{-12}\text{F}$）至毫法（$1\text{ mF}=10^{-3}\text{ F}$）级的。

电容的电路符号如图 4.2 所示，这里电压和电流取关联的参考方向。根据电流的定义——电荷移动的速率，有

$$i = \frac{\mathrm{d}q}{\mathrm{d}t} \tag{4.2}$$

将式（4.1）代入式（4.2），并考虑电容 C 一般为常数，与时间无关，因此电容的电流与电压关系可表示为

$$i = \frac{\mathrm{d}}{\mathrm{d}t}(Cv) = C\frac{\mathrm{d}v}{\mathrm{d}t} \tag{4.3}$$

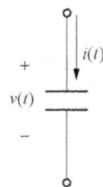

图 4.2　电容元件的电路符号，（电压、电流取关联参考方向）

Fig.4.2　The circuit symbol of capacitor (the voltage and the current have passive reference configuration)

由此可以看出，当电压增加时，有正向电流流过电容；若电压保持不变，没有电流流过，电容相当于开路；而当电压减小时，电流反向流过电容。

In previous chapters, we only considered three types of circuit elements—conductors, sources and resistors. In this chapter, we will study two additional circuit elements—capacitors and inductors. Unlike resistors that consume energy, capacitors and inductors are energy-storage elements; they store energy and feed it back to the circuit at a later time. Because of these two elements, the sudden removal or application of DC sources (or the sudden application of a signal, usually caused by switching) results in the time-varying voltages and currents. Such a phenomenon is called transient. Correspondingly, steady state refers to the state when there is no change occuring in the circuit structure or element parameters. Transient can be described as the transformation of a circuit from one steady state to another steady state. In this chapter the circuits we will discuss involve both capacitors and inductors. A circuit may work in transient or steady state,yet the principles of circuit analysis remain the same, and all the methods and principles we learnt in the previous chapters can be applied in the same manner.

4.1   Capacitance and Inductance

4.1.1   Capacitance

A capacitor is an ideal circuit element, which represents the ability to store energy in the form of electrical charge. A practical capacitor may take various forms, among which a simple and common form is that of a parallel-plate capacitor. It is composed of two sheets of conductor, which is usually metallic, separated by a thin layer of insulating material. The insulating material is called dielectric and can be comprised of air, Mylar, mica, electrolyte, or a variety of non-conduct-ing materials. Suppose the current is flowing downwards to the upper plate, as shown in Fig. 4.1. What actually happens is electrons flow upward and accumulate on the lower plate, resulting in the lower plate accumulating a negative charge. Therefore, an electric field is produced in the dielectric.

The charge stored by a capacitor is proportional to the voltage between the plates

$$q = Cv \tag{4.1}$$

where C denotes the capacitance, whose unit is farads (F). A farad is a very large amount of capacitance. In most cases the capacitances we use have values ranging from picofarads (1 pF=10^{-12} F) to millifarads (1 mF=10^{-3} F).

The circuit symbol for capacitor is shown in Fig. 4.2, where the voltage and the current are of passive reference configuration. Since current is defined as the rate of flow of charge per unit time , we have

$$i = \frac{\mathrm{d}q}{\mathrm{d}t} \tag{4.2}$$

Substituting (4.1) into (4.2) and noting that capacitance C is not a function of time, the relationship between current and voltage becomes

$$i = \frac{\mathrm{d}}{\mathrm{d}t}(Cv) = C\frac{\mathrm{d}v}{\mathrm{d}t} \tag{4.3}$$

Equation (4.3) shows that as the voltage increases, a positive current flows into the positive polarity of the capacitor; if the voltage remains constant, the current is zero and the capacitance acts as an open circuit; when the voltage decreases, a negative current will flow into the capacitor.

显然，当电压与电流取非关联的参考方向时，二者关系为

$$i = -C \frac{\mathrm{d}v}{\mathrm{d}t} \tag{4.4}$$

以上为通过电压求电容电流的方法。下面讨论当流过电容的电流为已知的情况。首先计算电荷，即某段时间内电流的积分

$$q(t) = \int_{t_0}^{t} i(t)\mathrm{d}t + q(t_0) \tag{4.5}$$

其中初始时刻 t_0 的电量 $q(t_0)$ 为已知。将式（4.5）代入式（4.1），得

$$v(t) = \frac{1}{C} \int_{t_0}^{t} i(t)\mathrm{d}t + \frac{q(t_0)}{C} \tag{4.6}$$

其中 $\frac{q(t_0)}{C}$ 为初始时刻电压，以 $v(t_0)$ 表示，则式（4.6）可写为

$$v(t) = \frac{1}{C} \int_{t_0}^{t} i(t)\mathrm{d}t + v(t_0) \tag{4.7}$$

可见电容电压不仅与 $t_0 \sim t$ 时刻的电流有关，也与电压在 t_0 时刻的初始值有关。

在给定电压与电流关联参考方向的情况下，电容吸收的功率为

$$p(t) = v(t)i(t) = Cv \frac{\mathrm{d}v}{\mathrm{d}t} \tag{4.8}$$

$t_0 \sim t$ 时刻电容吸收的能量为

$$w(t) = \int_{t_0}^{t} p(t)\mathrm{d}t = \int_{t_0}^{t} Cv \frac{\mathrm{d}v}{\mathrm{d}t}\mathrm{d}t = C \int_{v_0}^{v} v\mathrm{d}t$$
$$= \frac{1}{2}Cv^2(t) - \frac{1}{2}Cv^2(t_0) \tag{4.9}$$

当初始条件为 $v(t_0) = 0$ 时，式（4.9）变为

$$w(t) = \frac{1}{2}Cv^2(t) \tag{4.10}$$

再考虑到电容电压与电容存储电荷之间的关系，有

$$w(t) = \frac{1}{2}v(t)q(t) = \frac{q^2(t)}{2C} \tag{4.11}$$

4.1.2 电容的串并联

电容的串联或并联条件下的等效电容可以通过应用 KCL 和 KVL 计算得到。以电容并联为例，如图 4.3 所示，三个电容两端的电压相同，则其电流分别为

$$i_1 = C_1 \frac{\mathrm{d}v}{\mathrm{d}t} \qquad i_2 = C_2 \frac{\mathrm{d}v}{\mathrm{d}t} \qquad i_3 = C_3 \frac{\mathrm{d}v}{\mathrm{d}t} \tag{4.12}$$

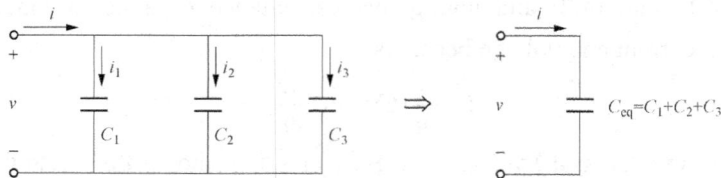

图 4.3　三个电容并联及其等效电容

Fig. 4.3　Three capacitors connected in parallel and the equivalent capacitance

If the reference polarity of voltage and the reference direction of current are against the passive reference configuration, the relationship between current and voltage becomes

$$i = -C\frac{dv}{dt} \tag{4.4}$$

Now that we know how to determine the current given the voltage, next we discuss how to express voltage in terms of current. First we compute the charge, which is the integration of the current over a period of time

$$q(t) = \int_{t_0}^{t} i(t)dt + q(t_0) \tag{4.5}$$

The initial charge $q(t_0)$ at some initial time t_0 is known. Substituting the above equation into (4.1) gives

$$v(t) = \frac{1}{C}\int_{t_0}^{t} i(t)dt + \frac{q(t_0)}{C} \tag{4.6}$$

where $\frac{q(t_0)}{C}$ is the initial voltage across the capacitor. Denoted by $v(t_0)$, (4.6) becomes

$$v(t) = \frac{1}{C}\int_{t_0}^{t} i(t)dt + v(t_0) \tag{4.7}$$

Apparently, the voltage of the capacitor is not only determined by the current from time t_0 to time t, but it also depends on the initial voltage.

Under the passive reference configuration, the power absorbed by a capacitor is

$$p(t) = v(t)i(t) = Cv\frac{dv}{dt} \tag{4.8}$$

Therefore, the the amount of energy delivered to the capacitor from time t_0 to time t is

$$w(t) = \int_{t_0}^{t} p(t)dt = \int_{t_0}^{t} Cv\frac{dv}{dt}dt = C\int_{v_0}^{v} vdt$$
$$= \frac{1}{2}Cv^2(t) - \frac{1}{2}Cv^2(t_0) \tag{4.9}$$

If initially we have $v(t_0) = 0$, the above equation becomes

$$w(t) = \frac{1}{2}Cv^2(t) \tag{4.10}$$

Taking the relationship between the voltage and the stored charge, we have

$$w(t) = \frac{1}{2}v(t)q(t) = \frac{q^2(t)}{2C} \tag{4.11}$$

4.1.2　Capacitors in series and parallel

The equivalent capacitance of capacitors connected in series or parallel can be calculated by applying KCL and KVL. Measuring the capacitance of a system in parallel for example. As shown in Fig. 4.3, the voltage across each capacitor is the same, and their corresponding currents can be obtained from

$$i_1 = C_1\frac{dv}{dt} \qquad i_2 = C_2\frac{dv}{dt} \qquad i_3 = C_3\frac{dv}{dt} \tag{4.12}$$

在顶部节点处应用 KCL，得

$$i = i_1 + i_2 + i_3 \tag{4.13}$$

将式（4.12）中的三个电流表达式代入式（4.13），可得

$$i = C_1 \frac{\mathrm{d}v}{\mathrm{d}t} + C_2 \frac{\mathrm{d}v}{\mathrm{d}t} + C_3 \frac{\mathrm{d}v}{\mathrm{d}t} + (C_1 + C_2 + C_3) \frac{\mathrm{d}v}{\mathrm{d}t} \tag{4.14}$$

另一方面，若将三个电容之和表示为 C_{eq}，则根据定义有

$$i = C_{\mathrm{eq}} \frac{\mathrm{d}v}{\mathrm{d}t} \tag{4.15}$$

对比式（4.14）与式（4.15），可知

$$C_{\mathrm{eq}} = C_1 + C_2 + C_3 \tag{4.16}$$

即并联电容的等效电容等于各电容之和。

串联电容的等效电容可以通过类似的方式推导出来。以图 4.4 所示的三个电容串联为例，其等效电容为

$$C_{\mathrm{eq}} = \frac{1}{1/C_1 + 1/C_2 + 1/C_3} \tag{4.17}$$

即串联电容的等效电容为各电容倒数之和的倒数。可见电容的串并联规则与电阻正好相反。请读者自行推导上述公式。求并联电容等效电容的方法与电阻串联相同，求串联电容等效电容的方法与电阻并联相同。

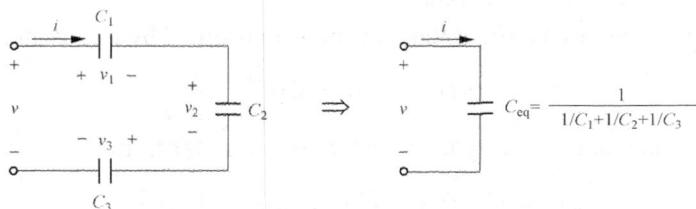

图 4.4　三个电容串联及其等效电容

Fig. 4.4　Three capacitors connected in series and the equivalent capacitance

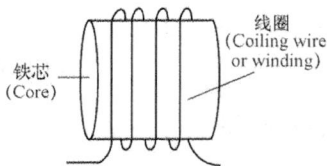

图 4.5　电感示意图

Fig. 4.5　A typical inductor

4.1.3　电感

电感由导线在某种材料上绕制而成，如图 4.5 所示，这样的导线称为线圈，常见的电感有空心的和带铁芯的。当有变化的电流通过线圈时会产生磁场，若所产生的磁场随时间变化，则在线圈中有感应电压产生。对于理想电感，感应电压与电流的变化率呈正比。

电感元件的电路符号如图 4.6 所示，这里电压与电流取关联参考方向，则电感的电压与电流间关系为

$$v(t) = L \frac{\mathrm{d}i}{\mathrm{d}t} \tag{4.18}$$

其中比例系数 L 为该电感元件的电感，单位为亨利（简称亨，H）。当电压与电流取非关联的参考方向时，其关系为

$$v(t) = -L \frac{\mathrm{d}i}{\mathrm{d}t} \tag{4.19}$$

Applying KCL at the top node of the circuit, we have

$$i = i_1 + i_2 + i_3 \tag{4.13}$$

Substituting the expressions in (4.12) into (4.13), we get

$$i = C_1 \frac{dv}{dt} + C_2 \frac{dv}{dt} + C_3 \frac{dv}{dt} = (C_1 + C_2 + C_3) \frac{dv}{dt} \tag{4.14}$$

On the other hand, denoting the equivalent capacitance by C_{eq}, we have the following expression by definition

$$i = C_{eq} \frac{dv}{dt} \tag{4.15}$$

Comparing (4.14) and (4.15), the following equation can be obtained

$$C_{eq} = C_1 + C_2 + C_3 \tag{4.16}$$

That is to say, the equivalent capacitance of parallel-connected capactiors is the sum of the individual capacitances.

We can apply the same procedure to calculate the equivalent capacitance of capactiors connected in series. Take the case shown in Fig. 4.4 for example, where three capacitors are combined in series. The equivalent capacitance is

$$C_{eq} = \frac{1}{1/C_1 + 1/C_2 + 1/C_3} \tag{4.17}$$

That is to say, the equivalent capacitance of series-connected capacitors is the reciprocal of the sum of the reciprocals of the individual capacitances. Apparently, the rules for series/parallel combinations of capacitances are the opposite to that for resistances. We leave it to the readers to derive this equation. To calculate the equivalent capacitance of series/parallel-conncted capacitors, we follow the same strategy as we calculate the equivalent resistance of parallel/series-connected resiators.

4.1.3　Inductance

Inductors are manufactured by coiling a wire around some type of material, usually air or iron, as shown in Fig. 4.5. Such a wire is called a winding. As a time-varying current flows through the winding, it creates a magnetic field. If the magnetic field also changes with time, a voltage is induced across the inductor. For an ideal inductor, the induced voltage is proportional to the rate of the change of the current per unit time.

The circuit symbol of an inductor is given in Fig. 4.6, where the voltage and the current have the passive reference configuration. The relationship between the voltage and the current is

$$v(t) = L \frac{di}{dt} \tag{4.18}$$

The constant of proportionality L is called inductance, and its unit is Henry, short for H. If the voltage and the current are against the passive reference configuration, their relationship is defined as

$$v(t) = -L \frac{di}{dt} \tag{4.19}$$

以上为通过电流求电感电压的方法，下面讨论当已知初始电流 $i(t_0)$ 和电感电压，求电感电流的情况。对式（4.18）两端求从 t_0 至 t 的积分，有

$$i(t) = \frac{1}{L}\int_{t_0}^{t} v(t)\mathrm{d}t + i(t_0) \tag{4.20}$$

从 t_0 至 t 时刻电感吸收的能量为

$$w(t) = \int_{t_0}^{t} p(t)\mathrm{d}t = \int_{t_0}^{t} Li\frac{\mathrm{d}i}{\mathrm{d}t}\mathrm{d}t = \int_{i(t_0)}^{i(t)} Li\,\mathrm{d}i \tag{4.21}$$

若初始条件为 $i(t_0)=0$，则有

$$w(t) = \frac{1}{2}Li^2(t) \tag{4.22}$$

4.1.4　电感的串并联

求串并联电感等效电感的思路与求串并联电阻与电容的方法相同，即应用 KCL 和 KVL。以图 4.7 所示三个电感串联为例，应用 KVL 可得

$$v = v_1 + v_2 + v_3 \tag{4.23}$$

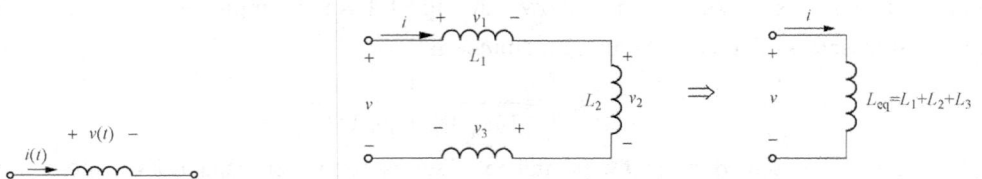

图 4.6　电感元件的电路符号

（电压与电流取关联参考方向）

Fig. 4.6　The circuit symbol of an inductor

图 4.7　三个电感串联及其等效电感

Fig. 4.7　Three inductors connected in series and the equivalent inductance

将电压-电流关系代入式（4.23），由于流过各电感的电流相同，可得

$$v = L_1\frac{\mathrm{d}i}{\mathrm{d}t} + L_2\frac{\mathrm{d}i}{\mathrm{d}t} + L_3\frac{\mathrm{d}i}{\mathrm{d}t} = (L_1 + L_2 + L_3)\frac{\mathrm{d}i}{\mathrm{d}t} = L_{eq}\frac{\mathrm{d}i}{\mathrm{d}t} \tag{4.24}$$

显然，串联电感的等效电感为

$$L_{eq} = L_1 + L_2 + L_3 \tag{4.25}$$

即串联电感的等效电感为各电感之和。

如图 4.8 所示，对三个并联电感应用 KCL，可得其等效电感为

$$L_{eq} = \frac{1}{1/L_1 + 1/L_2 + 1/L_3} \tag{4.26}$$

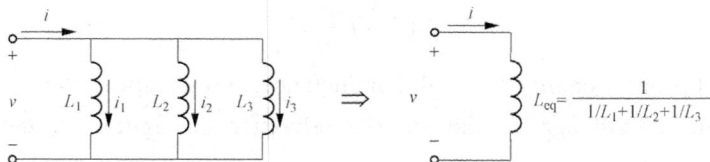

图 4.8　三个电感并联及其等效电感

Fig. 4.8　Three inductors connected in parallel and the equivalent inductance

We have just demonstrated how to calculate voltage from current. Next, let's consider how to find the current when the initial current $i(t_0)$ and voltage of the inductor are given. Perform integration for (4.18) from t_0 to t, we have

$$i(t) = \frac{1}{L} \int_{t_0}^{t} v(t)\mathrm{d}t + i(t_0) \tag{4.20}$$

The energy absorbed by the inductore from t_0 to t is

$$w(t) = \int_{t_0}^{t} p(t)\mathrm{d}t = \int_{t_0}^{t} Li\frac{\mathrm{d}i}{\mathrm{d}t}\mathrm{d}t = \int_{i(t_0)}^{i(t)} Li\mathrm{d}i \tag{4.21}$$

Suppose initially $i(t_0)=0$, we have

$$w(t) = \frac{1}{2}Li^2(t) \tag{4.22}$$

4.1.4　Inductors in series and parallel

To calculate the equivalent inductance of series/parallel combinations, we use the same strategy as we did for determining the equivalent resistance and capacitance of a circuit, which is to apply KCL and KVL. For the three inductors connected in series, as shown in Fig. 4.7, applying KVL gives us

$$v = v_1 + v_2 + v_3 \tag{4.23}$$

Replace the voltages by the voltage-current relationship. As the current for each inductor is the same, it results in

$$v = L_1\frac{\mathrm{d}i}{\mathrm{d}t} + L_2\frac{\mathrm{d}i}{\mathrm{d}t} + L_3\frac{\mathrm{d}i}{\mathrm{d}t} = (L_1 + L_2 + L_3)\frac{\mathrm{d}i}{\mathrm{d}t} = L_{eq}\frac{\mathrm{d}i}{\mathrm{d}t} \tag{4.24}$$

Consequently, the equivalent inductance of series-connected inductors is

$$L_{eq} = L_1 + L_2 + L_3 \tag{4.25}$$

That is to say, the equivalent inductance of series-connected inductors is the sum of the individual inductances.

As shown in Fig. 4.8, applying KCL to the three inductors connected in parallel, we find the equivalent inductance is

$$L_{eq} = \frac{1}{1/L_1 + 1/L_2 + 1/L_3} \tag{4.26}$$

Michael Faraday (22 September 1791 – 25 August 1867) was an English scientist who contributed to the fields of electromagnetism and electrochemistry. He was one of the most influential scientists in history. Faraday established the basis for the concept of the electromagnetic field and discovered the principles of electromagnetic induction and diamagnetism, and the laws of electrolysis. His inventions of electromagnetic rotary devices formed the foundation of electric motor technology, and it was largely due to his efforts that electricity became practical for use in technology. Faraday was also a chemist.

即并联电感的等效电感为各电感倒数之和的倒数。可见电感的串并联规则与电阻相同。

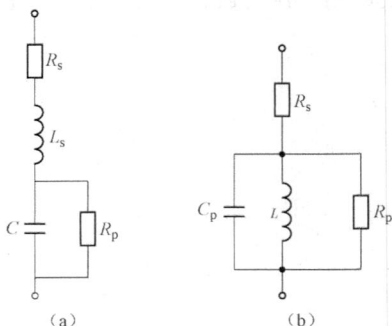

图 4.9 实际电容和电感更精确的模型

（a）实际电容器模型；（b）实际电感器模型

Fig. 4.9　The more accurate models for a real capacitor and a red inductor

(a) A model of a real capacitor; (b) A model of a real inductor

讨论： 实际中的电容与电感

在电路中电容器通常用其电容来表示，但是在实际中，更精确的电容器模型除电容 C 外，还应包括如下几部分：描述电容器极板阻值的电阻 R_s，它与电容串联；由于电流流经电容器而产生的电感 L_s，它也与电容串联；代表电介质阻值的电阻 R_p，它与电容并联。具体的电路图如图 4.9（a）所示。

同样地，实际中的电感器也不能仅用一个电感来表示。更精确的模型如图 4.8（b）所示，除电感外还包括如下几部分：电感绕组的电阻 R_s，它与电感串联；绕组间的电容 C_p，它与电感并联；以及反映铁芯损耗的电阻 R_p，它也与电感并联。

4.2 直流稳态分析

通过上节中对电容电感的电压电流关系的分析可以看出，对于直流电路而言，电容相当于开路，而电感相当于短路。由式（4.3）可知，当电容电压保持不变时，电容电流为 0，即相当于开路。由式（4.18）可知，当电感电流为恒定时，电感电压为 0，相当于短路。在这种条件下的电路分析称为直流稳态分析，基本分析方法为：首先将所有电容用开路代替，将所有电感以短路代替，余下的电路中仅含有电源和电阻，对其应用前面几章介绍过的电路分析方法即可。

【例 4.1】 如图 4.10（a）所示电路中，$i_s = 9\,\text{mA}$，$R_1 = 6\text{k}\Omega$，$R_2 = 3\text{k}\Omega$，$C = 2\mu\text{F}$，$L = 1\text{mH}$，求当开关闭合很久后电阻 R_1 两端电压。

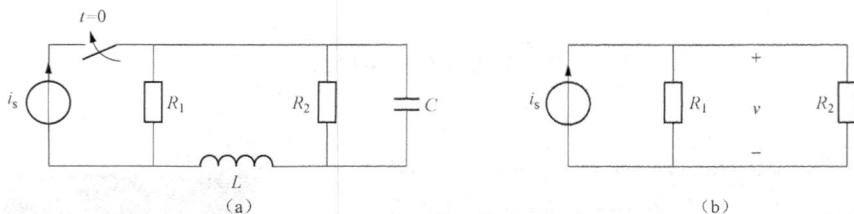

图 4.10　例 4.1 的电路图

（a）原电路；（b）稳态下的等效电路图

Fig. 4.10　Circuit diagram for Example 4.1

(a) Original circuit; (b) Equivalent circuit for the steady state

解 当开关闭合很久后，电路处于稳态，此时电感可视为短路，电容可视为开路，因此其等效电路如图 4.10（b）所示，电阻 R_1、R_2 并联，其电压为

$$v = \frac{R_1 R_2}{R_1 + R_2} \cdot i_s = \frac{6 \times 3}{6 + 3} \times 9 = 18\text{V}$$

According to the above equation, the equivalent inductance of parallel-connected inductors is the reciprocal of the sum of the reciprocals of the individual inductances. Apparently, the series/parallel combination rules for determining inductances are the same as those for resistances.

Discussion: Practical capacitors and inductors

In circuit analysis, we usually model a capacitor by its capacitance. However, in practice, a real capacitor cannot be represented simply by its capacitance alone. A more accurate circuit model for a real capacitor is given in Fig. 4.9(a). Here, in addition to the capacitor, C, we include a resistor R_s connected in series with it, which is due to the resistance of the capacitor's plates;we also include an inductor L_s connected in series with the capacitor, which appears because the current flowing through the capacitor creates a magnetic field. Finally we add a resistor R_p connected in parallel with the capacitor, which describes the resistance through the dielectric.

Likewise, a real inductor cannot simply be represented by its inductance. As shown in Fig. 4.9(b), a more realistic model also involves a series resistor R_s, which is caused by the resistance of the winding material; a parallel capacitor C_p, which reflects the capacitance between the layers of the winding; and a parallel resistor R_p, which describes the core loss.

4.2 DC Steady State Analysis

From the voltage-current relationships of capacitors and inductors, it can be seen that capacitor behaves as an open circuit for DC sources, while inductor acts as a short circuit. Equation (4.3) implies that when the voltage of a capacitor is constant, its current is zero; thus, it appears as an open circuit. From (4.18) we can tell that when the current of an inductor is constant, its voltage is zero; hence, it appears as a short circuit. Circuit analysis under such conditions with DC sources is called DC steady state analysis. We analyze such a circuit by executing the following steps: First, replace all the capacitors by open circuits and replace all the inductors by short circuits. The remaining circuit contains resistors and sources only. Then applying any suitable circuit analysis method,which we have learnt in the previous chapters we can solve the remaining circuit.

Example 4.1: Consider the circuit shown in Fig. 4.10(a), where $i_s = 9$ mA, $R_1 = 6\,\text{k}\Omega$, $R_2 = 3\,\text{k}\Omega$, $C = 2\,\mu\text{F}$, $L = 1\,\text{mH}$. Find the voltage of R_1 for $t \gg 0$.

Solution:

After the switch has been closed for a long time, the circuit works in steady state. Therefore, the inductor can be replaced by a short circuit and the capacitor by an open circuit. The equivalent circuit is shown in Fig. 4.10(b), where resistances R_1 and R_2 are connected in parallel. Thus, we have

$$v = \frac{R_1 R_2}{R_1 + R_2} i_s = \frac{6 \times 3}{6 + 3} \times 9 = 18 \text{ V}$$

4.3 一阶 RC 电路

由于储能元件的存在，当突然有直流电源切入或切出时（通常由开关控制），电路中有随时间变化的电压或电流产生，这一现象称为暂态。在这两节中将介绍两种基本暂态电路：一种电路中仅含有电阻和电容，称为一阶 RC 电路；另一种仅含有电阻和电感，称为一阶 RL 电路。与分析电阻电路相同，我们仍然应用 KCL 和 KVL 分析这两种电路。唯一的区别在于应用 KCL 和 KVL 分析电阻电路时得到代数方程，而用它们分析一阶 RC 和 RL 电路则得到微分方程。

如图 4.11（a）所示，电路处于稳态，开关打开，电容已充电，有初始电压 V_0。在 $t=0$ 时刻将开关闭合，会有电流流过电阻，电容放电。设电容电压为 $v_C(t)$，对开关闭合后的电路应用 KCL，有

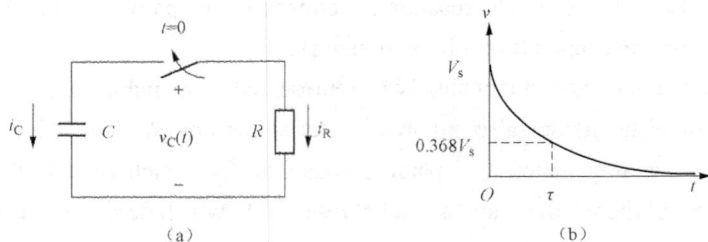

图 4.11 一阶电容放电 RC 电路

（a）电路图；（b）电容电压时间曲线

Fig.4.11 A first-order RC circuit with a capacitor being discharged

(a) Circuit diagram; (b) Voltage versus time for capacitance voltage $v_C(t)$

$$C\frac{\mathrm{d}v_C(t)}{\mathrm{d}t} + \frac{v_C(t)}{R} = 0 \tag{4.27}$$

整理得到

$$RC\frac{\mathrm{d}v_C(t)}{\mathrm{d}t} + v_C(t) = 0 \tag{4.28}$$

显然这是一个关于 $v_C(t)$ 的一阶微分方程，且等号右边为 0，则其解必然具有如下指数形式

$$v_C(t) = Ke^{st} \tag{4.29}$$

只需要确定系数 K 和 s 即可。将式（4.29）代入式（4.28）可得

$$RCKse^{st} + Ke^{st} = 0 \tag{4.30}$$

从式（4.30）中可以求解出系数 s

$$s = -\frac{1}{RC} \tag{4.31}$$

将 s 代入式（4.29）中，有

$$v_C(t) = Ke^{-\frac{t}{RC}} \tag{4.32}$$

下面继续求解系数 K。此时需要用到初始条件 $v_C(0_+) = V_0$，即当开关刚刚闭合时，电容电压尚未发生变化，仍为初始值 V_0。将此初始条件代入式（4.32）中，可得 $K = V_0$。将系数 K 与 s

4.3   First–Order RC Circuits

Because of the existence of energy storage elements, sudden application or removal of a DC source, usually controlled by a switch, will result in the generation of time-varying currents or voltages in the circuit. Such a phenomenon is called a transient. In the following two sections, we will introduce two types of transient circuits: one circuit consists of a resistor and a capacitor, called the first-order RC circuit; and the other one comprises a resistor and an inductor, called the first-order RL circuit. As we did for resistive circuits, we also apply KCL and KVL to solve for RC and RL circuits. The only difference lies in that applying KCL and KVL to resistive circuits results in algebraic equations, whereas applying them to RC and RL circuits yields differential equations.

For the circuit shown in Fig. 4.11(a), it is in the steady state with the switch open for a long time. The capacitor is charged to an initial voltage V_0. The switch closes at time instant $t=0$, so current flows through the resistor and the capacitor is discharged. Denote the voltage across the capacitor by $v_C(t)$. Applying KCL to the circuit with switch closed, we have

$$C\frac{dv_C(t)}{dt} + \frac{v_C(t)}{R} = 0 \tag{4.27}$$

Rearranging the above equation, we get

$$RC\frac{dv_C(t)}{dt} + v_C(t) = 0 \tag{4.28}$$

This is a first-order differential equation about $v_C(t)$ whose right-hand side is zero. Therefore, its solution must have the form of

$$v_C(t) = Ke^{st} \tag{4.29}$$

And we only need to determine coefficients K and s. Substituting (4.29) into (4.28), we have

$$RCKse^{st} + Ke^{st} = 0 \tag{4.30}$$

from which we can easily solve for s

$$s = -\frac{1}{RC} \tag{4.31}$$

Replacing s into (4.29) yields

$$v_C(t) = Ke^{-\frac{t}{RC}} \tag{4.32}$$

Next, we solve for coefficient K. This time we use the initial condition of $v_C(0_+) = V_0$. When the switch just closes, the voltage of the capacitor does not change yet, so it remains at V_0. Substituting this initial condition into (4.32) gives us $K = V_0$. Substituting

代入式（4.29）中，可以得到电容电压随时间变化的表达式

$$v_C(t) = V_0 e^{-\frac{t}{RC}}$$ （4.33）

电容电压的时间曲线如图 4.11（b）所示。通常我们记 $\tau = RC$ 为一阶 RC 电路的时间常数。在一个时间常数内，电容电压将下降至原来的 $e^{-1} \approx 0.368$。

　　以上讨论的是电容放电过程，下面对电容充电过程进行暂态分析。如图 4.12（a）所示，在 $t=0$ 时刻之前开关打开，电路处于稳态，电容两端电压为 0，即 $v_C(0_-)=0$。在 $t=0$ 时刻开关突然闭合，直流电压源 V_s 接入电路，对电容充电。但电容电压不会突然变化，即 $v_C(0_+)=0$。在充电过程中对电路应用 KVL，有

$$V_s = i(t)R + v_C(t) = RC\frac{dv_C(t)}{dt} + v_C(t)$$ （4.34）

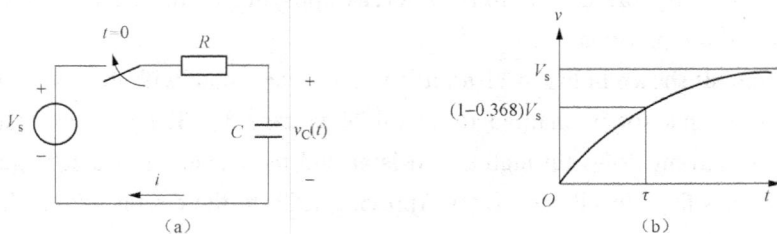

图 4.12　一阶电容充电 RC 电路

（a）电路图；（b）电容电压时间曲线

Fig. 4.12　A first-order RC circuit with a capacitor being charged

(a) Circuit diagram; (b) Voltage versus time for capacitance voltage $v_C(t)$

对于这个等号左边为非零常数的一阶微分方程，其解必然具有如下形式

$$v_C(t) = K_1 + K_2 e^{st}$$ （4.35）

将式（4.35）代入式（4.34）中，得到 $s = -\dfrac{1}{RC}$，$K_1 = V_s$。将系数 s 和 K_1 与初始条件 $v_C(0_+) = 0$ 一起代入式（4.35），可得 $K_2 = -V_s$。将三个系数均代入式（4.35），得到电容电压表达式为

$$v_C(t) = V_s - V_s e^{-\frac{t}{RC}}$$ （4.36）

其中等号右边第一项称为稳态响应，等号右边第二项称为暂态响应。时间常数 $\tau = RC$ 的概念仍适用。电容电压的时间曲线如图 4.12（b）所示，开关闭合后，电容电压按指数逐渐增加至电源电压 V_s，而非突变至 V_s。

4.4　一阶 RL 电路

　　一阶 RL 电路主要由直流电源、电阻和电感组成，本节讨论如何对一阶 RL 电路进行分析，采用的方法与上节所述的分析一阶 RC 电路的方法相同。以图 4.13（a）所示一阶 RL 电路为例，在 $t<0$ 时开关断开，电路处于稳态；在 $t=0$ 时开关闭合，电路中有电流流过，设其为 $i(t)$。可列出电路的 KVL 方程如下

$$V_s = Ri(t) + v_L(t) = Ri(t) + L\frac{di(t)}{dt}$$ （4.37）

coefficient K and s into (4.29), the solution for the time-varying voltage is

$$v_C(t) = V_0 e^{-\frac{t}{RC}} \tag{4.33}$$

A plot of the voltage against time is shown in Fig. 4.11(b). Usually $\tau = RC$ is defined as the time constant of the first-order RC circuit. In one time constant, the voltage drops to $e^{-1} \approx 0.368$ of its original value.

Now let's consider what happens when a capacitor is being charged. As shown in Fig. 4.12(a), the switch is open before $t=0$, the circuit is in steady state, and the voltage across the capacitor is $v_C(0_-)0$. At time instant $t=0$, the switch suddenly closes, introducing the DC voltage source V_s into the circuit, which charges the capacitor. Yet the voltage of the capacitor does not change instantly; that is to say $v_C(0_+) = 0$. Applying KVL to the charging circuit, we have

$$V_s = i(t)R + v_C(t) = RC\frac{dv_C(t)}{dt} + v_C(t) \tag{4.34}$$

For such a differential equation whose left-hand side is a non-zero constant, its solution must follow the pattern of

$$v_C(t) = K_1 + K_2 e^{st} \tag{4.35}$$

Substituting (4.35) into (4.34) gives $s = -\dfrac{1}{RC}$ and $K_1 = V_s$. Using coefficients s and K_1, and the initial condition $v_C(0_+) = 0$ to solve (4.35), we obtain $K_2 = -V_s$. Substituting the three coefficients into (4.35), the time-varying voltage of the capacitor is solved as

$$v_C(t) = V_s - V_s e^{-\frac{t}{RC}} \tag{4.36}$$

in which the first term on the right-hand side is the steady-state response and the second term corresponds to the transient response. The time constant is also defined as $\tau = RC$. A plot of voltage $v_C(t)$ is illustrated in Fig. 4.12(b). After the switch is closed, instead of jumping to V_s, the voltage of the capacitor increases exponentially to source voltage V_s.

4.4   First–Order RL Circuits

In this section, we consider first-order RL circuits, which consist of DC sources, resistors, and inductors. In order to solve such a circuit, we apply a very similar method as we used in the previous section to solve for RC circuits. Consider the first-order RL circuit shown in Fig. 4.13(a). The switch is open for $t<0$ and the circuit is in steady state. At $t = 0$ the switch closes, resulting a current flowing through the inductor. Denote the current by $i(t)$ and we can obtain the KVL equation as follows

$$V_s = Ri(t) + v_L(t) = Ri(t) + L\frac{di(t)}{dt} \tag{4.37}$$

显然这也是一个一阶微分方程，其解具有如下形式

$$i(t) = K_1 + K_2 e^{st} \tag{4.38}$$

将式（4.38）代入式（4.37）中，可以解得 $K_1 = V_s / R$ 和 $s = -R/L$。再将这两个系数代入式（4.37），同时考虑初始条件 $i(0_+) = 0$，有

$$0 = K_1 + K_2 \Rightarrow K_2 = -K_1 = -V_s / R \tag{4.39}$$

则开关闭合后电路中的电流为

$$i(t) = \frac{V_s}{R} - \frac{V_s}{R} e^{-Rt/L} \tag{4.40}$$

定义时间常数 $\tau = L/R$，则上式可表示为如下形式

$$i(t) = \frac{V_s}{R} - \frac{V_s}{R} e^{-t/\tau} \tag{4.41}$$

该电路的电流—时间曲线如图 4.13（b）所示，电流按指数从 0 增加至 V_s/R，即达到第二个稳态时的电流值（此时电感相当于短路）。

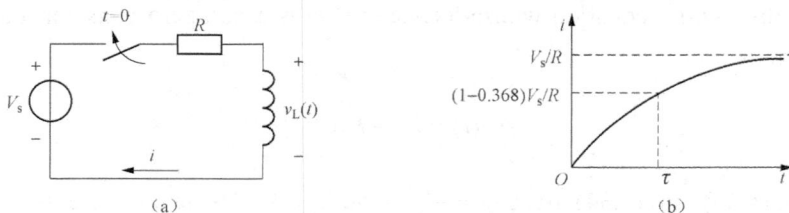

图 4.13　一阶 RL 电路

（a）电路图；（b）电感电流—时间曲线

Fig. 4.13　A first-order RL circuit

(a) Circuit diagram; (b) Current versus time for inductance current

4.5　三要素分析法

综合 4.3 节与 4.4 节可以看出，一阶微分电路的分析步骤为：

（1）应用 KCL 或 KVL 列出电路方程；

（2）假设方程的解具有 $K_1 + K_2 e^{st}$ 的形式；

（3）将该解代入电路方程求解系数 K_1 和 s；

（4）应用初始条件求解系数 K_2；

（5）列出解的最终表达式。

分析系数 K_1 和 K_2 的求解过程可以发现，K_1 为所求电压或电流在电路到达第二个稳态时的稳态解，而 K_2 为该电压或电流的初始值与稳态解之差，即

$$v(t) = v(\infty) + [v(0_+) - v(\infty)] e^{-t/\tau} \tag{4.42}$$

其中 $v(0_+)$ 为开关动作后瞬间的电压值，等于电路处于第一个稳态时的电压值，$v(\infty)$ 是电路处于第二个稳态时的电压值，τ 为时间常数。电流有同样的表达形式。也就是说，只要知道了初始值、稳态解和时间常数这三个要素，即可得出电路在直流激励下的电压或电流的表达式，这就是三要素法。

For such a first-order differential equation, it must have a solution in the form of

$$i(t) = K_1 + K_2 e^{st} \tag{4.38}$$

Substituting (4.38) into (4.37), we get $K_1 = V_s / R$ and $s = -R/L$. Replacing the two coefficients into (4.37) and engaging the initial current $i(0_+) = 0$, we obtain

$$0 = K_1 + K_2 \Rightarrow K_2 = -K_1 = -V_s / R \tag{4.39}$$

Thus, after the switch is closed the current can be expressed by

$$i(t) = \frac{V_s}{R} - \frac{V_s}{R} e^{-Rt/L} \tag{4.40}$$

Define the time constant as $\tau = L/R$ and the above equation becomes

$$i(t) = \frac{V_s}{R} - \frac{V_s}{R} e^{-t/\tau} \tag{4.41}$$

A plot of the current against time is demonstrated in Fig. 4.13(b), where the current increases exponentially from 0 to V_s/R, which is the current value of the second steady state when the inductance acts as a short circuit.

4.5 The Three−element Analysis Method

As it can be concluded from Sections 4.3 and 4.4, the analysis of a first-order differential circuit takes the following steps:

(1) Apply KCL and KVL to write the circuit equation;

(2) Assume the solution to the equation has the form of $K_1 + K_2 e^{st}$;

(3) Substitute the solution to the circuit equation to find out coefficients K_1 and s;

(4) Use the initial conditions to determine the value of coefficient K_2;

(5) Write the final solution.

From the way we solve for K_1 and K_2 it can be found that K_1 is the steady-state solution when the circuit reaches the second steady state, while K_2 is the difference between the initial condition and the steady-state solution, i.e.

$$v(t) = v(\infty) + [v(0_+) - v(\infty)] e^{-t/\tau} \tag{4.42}$$

where $v(0_+)$ represents the voltage immediately after the switch opens or closes, which equals to the voltage of the first steady state; $v(\infty)$ represents the voltage of the second steady state, and τ is the time constant. We can write a similar expression in terms of current. That is to say, as long as we know the variables of initial condition, steady-state solution, and time constant, we can easily write the expression of current or voltage with DC excitation. This is basically the three-element analysis method.

✎　**讨论：**　非线性电路的稳态与暂态、零状态响应与零输入响应

　　分析各类元件的电压—电流关系可知，电容和电感与电阻的不同之处在于：电阻的伏安特性是线性的，$v=Ri$，而电容和电感的伏安特性是非线性的，$i=C\dfrac{\mathrm{d}v}{\mathrm{d}t}$，$v=L\dfrac{\mathrm{d}i}{\mathrm{d}t}$。由于这种微分关系的存在，约束了电容电压和电感电流均为连续函数，即它们的值不能突然发生变化。事实上，它们是按指数规律变化的，这一过程即为暂态。

　　考虑式（4.42）中电压的三要素法表达式。等式左边称为一阶电路在 $t \geqslant 0$ 时的全响应，等式右边第一项称为稳态响应，第二项为暂态响应，它随时间增加而按指数规律衰减至零。

　　将式（4.42）改写为如下形式

$$v(t) = v(0_+)\mathrm{e}^{-t/\tau} + v(\infty)(1 - \mathrm{e}^{-t/\tau}) \tag{4.43}$$

则可从另一角度分析电路的全响应。上式右边第一项反映了当电源电压 $V_s=0$ 时电路的响应，称为零输入响应；右边第二项为当初始电压 $v(0_+)=0$ 时电路的响应，称为零状态响应；零输入响应与零状态响应之和即为电路的全响应。可以从叠加原理的角度思考这一现象。因为电容为储能元件，可以姑且看为另一个电源，则零输入响应是直流电源置零时的响应，而零状态响应是电容初始电压置零时的响应，电路的全响应为二者的叠加。

Discussion: Steady state and transient of non-linear circuits; zero state response and zero input response

From the voltage-current relationships of the three circuit elements, it can be found that the difference between capacitor/inductor and resistor is that the voltage-current relationship of resistor is linear: $v=Ri$, whereas that of capacitor and inductor is non-linear: $i = C\dfrac{dv}{dt}$, $v = L\dfrac{di}{dt}$. Because of the differential relationship, capacitor voltage and inductor current have to be continuous functions; thus, they cannot change abruptly. As a matter of fact, they change exponentially. Such a process is called transient.

Consider the three-element analysis in (4.42). $v(t)$ on the left-hand side of the equation is called the total response of the first-order circuit when $t \geqslant 0$. The first term on the right-hand side is the steady-state response, and the second term on the right is the transient response, which decreases to zero exponentially as time passes.

We can rewrite (4.42) into the following form

$$v(t) = v(0_+)e^{-t/\tau} + v(\infty)(1 - e^{-t/\tau})\qquad(4.43)$$

and view the problem from another prospective. The first term on the right-hand side is the response of the circuit when source voltage is zero ($V_s=0$), which is called the zero input response. The second term is the response of the circuit when the initial voltage is zero [$v(0_+)=0$], and it is called the zero state response. The sum of the zero input response and the zero state response correspond to the total response. It can be explained by the superposition principle. As an energy-storage element, the capacitor can be considered as a source. As such, the zero input response is the response when we zero the DC source, and the zero state response is the response when we set the initial voltage to zero. And the total response is the superposition of the two.

▶ Problems

P4.1   Calculate the charge stored on a 5 pF capacitor with 10 V voltage across it. Calculate the energy stored in this capacitor.

P4.2   Suppose the voltage across a 100 μF capacitor has the waveform as shown in Fig. P4.1. Determine the current through it.

P4.3   Find the equivalent capacitance for each of the circuits shown in Fig. P4.2.

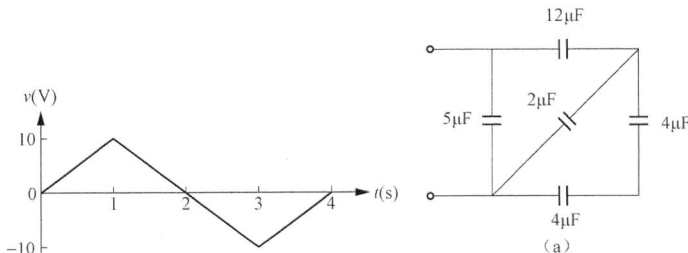

Fig. P4.1

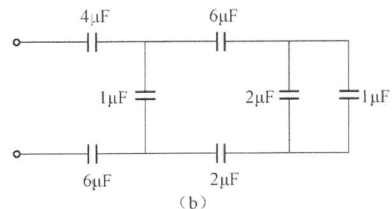

Fig. P4.2

91

P4.4　The current through a 0.1 H inductor is

$$i(t) = \begin{cases} 0 & t<0 \\ te^{-2t} & t\geqslant0 \end{cases}$$

Find the voltage across it.

P4.5　Find the equivalent inductance for each of the circuits shown in Fig. P4.3.

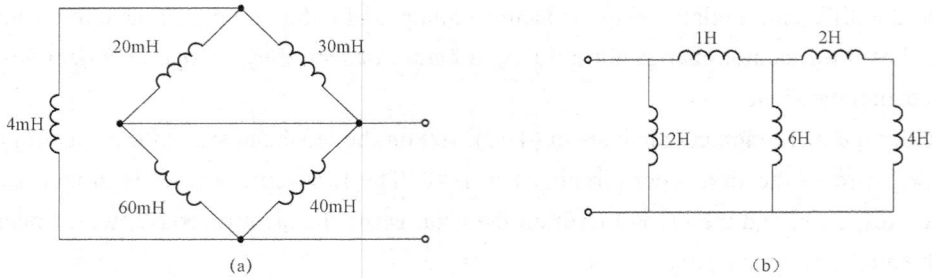

(a)　　　　　　　　　　　　　　　　　　(b)

Fig. P4.3

P4.6　Suppose the circuit in Fig. P4.4, with $i_s = 5\,\text{A}$, $R_1 = 2\,\Omega$, $C = 160\,\mu\text{F}$, $L = 4\,\text{mH}$, operates in DC steady state. Find the value of R so that the energy stored in the capacitor is the same as the energy stored in the inductor.

P4.7　The circuit shown in Fig. P4.5 is operating in DC steady state with the switch closed prior to $t = 0$. Find the expression for $v_C(t)$ when $t<0$ and $t\geqslant0$. Calculate the current i_C when $t=1\text{s}$.

P4.8　For the circuit shown in Fig. P4.6, assuming that $i(0)=10\,\text{A}$, calculate $i(t)$ and $i_x(t)$.

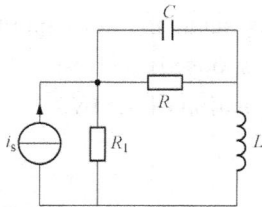
Fig. P4.4

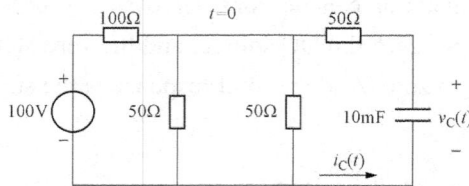
Fig. P4.5

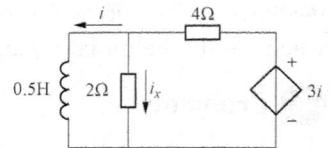
Fig. P4.6

正弦电路稳态分析
Steady–state
Sinusoidal Analysis

在上一章中学习了以电容、电感、电阻为负载的电路在直流电源激励下的表现与分析方法，在本章将考虑正弦交流电源激励下的电路分析方法。虽然交流电的波形可以是三角波、方波等，但人们通常说到交流电都是指正弦交流电，它在我们的生产生活中具有非常广泛的应用。

5.1　正弦电流与电压

图 5.1 所示为一正弦交流电压的波形图，其电压值随时间周期性变化且有正有负，满足如下条件

$$v(t) = V_m \cos(\omega t + \theta) \tag{5.1}$$

式中，V_m 为电压幅值；ω 是角频率，单位为弧度/秒（rad/s）；θ 为相角，单位为弧度（rad）或度（°），弧度和角度的换算关系为 $\varphi = 180°\theta/\pi$；周期信号在时间 T 内的波形是相同的，则时间 T 称为信号的周期；频率 f 定义为每秒内出现的周期数，$f = 1/T$，单位为赫兹（Hz），与角频率的关系是 $\omega = 2\pi f$。

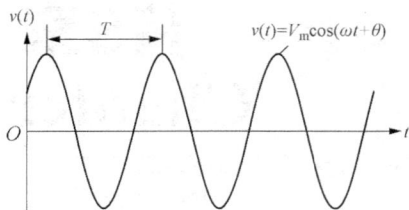

图 5.1　正弦交流电压波形图

Fig. 5.1　The waveform of a sinusoidal voltage

正弦电流的表达式通常有如下形式

$$i(t) = I_m \cos(\omega t + \theta) \tag{5.2}$$

在交流电路分析中，一般不区分正弦和余弦，为统一，本书中交流电都用余弦函数表达。若遇到以正弦表达的电流或电压，则应将其转换为余弦形式

$$\sin(\omega t + \theta) = \cos(\omega t + \theta - 90°) \tag{5.3}$$

在交流电路分析中也经常用到电压和电流的有效值，也称为均方根值，定义分别为

$$V_{rms} = \sqrt{\frac{1}{T} \int_0^T v^2(t)\,dt}$$

$$I_{rms} = \sqrt{\frac{1}{T} \int_0^T i^2(t)\,dt}$$

对于正弦电压和电流，通过三角函数计算可知

$$V_{rms} = \frac{V_m}{\sqrt{2}} \tag{5.4}$$

$$I_{rms} = \frac{I_m}{\sqrt{2}} \tag{5.5}$$

5.2　相量法

5.2.1　复数

相量法是一种以简单的相量形式代替复杂的余弦函数形式来方便地表达正弦信号并进行电路分析的方法。在介绍相量法之前首先需要回顾一下复数的基本知识。

假设有复数 $z = x + jy$，其中 x 为实部，y 为虚部，$j = \sqrt{-1}$ 为虚数单位（数学中虚数单位通常用字母 i 表示，但是电路中为了与电流区分，用 j 来表示）；这是复数的代数形式。将 z 画到如图 5.2 所示的复平面上，可以得

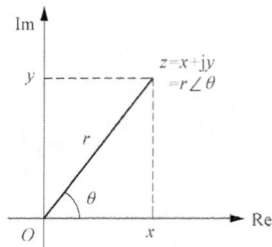

图 5.2　复数 $z = x + jy$ 在复平面上的示意图

Fig. 5.2　An illustration of a complex number $z = x + jy$ on the complex plane

In the previous chapter, we learnt about the behaviors of circuits powered by DC sources whose loads are composed of capacitors, inductors, and resistors, and how to analyze them. In this chapter, we will discuss how to analyze circuits powered by AC sources. Generally, AC sources can be triangle waves, square waves, etc; yet in most cases, AC sources refer to sinusoidal sources as they have vast application in our everyday life.

5.1   Sinusoidal Currents and Voltages

A typical sinusoidal voltage is shown in Fig. 5.1, whose magnitude varies with time periodically and can be either positive or negative. It can be expressed by the following function

$$v(t) = V_m \cos(\omega t + \theta) \tag{5.1}$$

where V_m is the amplitude of the voltage; ω is the angular frequency whose unit is radian per second (rad/s); θ represents the phase angle measured in the unit of radian or degree, the conversion of one to the other follows $\varphi = 180°\theta / \pi$; a periodic signal repeats a certain pattern in a time interval of T, which is called the period of the signal; frequency f is defined as the number of periods appear with in an interval of one second, $f = 1/T$, and its unit is Hertz (Hz); the relationship between frequency and angular frequency is $\omega = 2\pi f$.

A sinusoidal current is usually expressed as

$$i(t) = I_m \cos(\omega t + \theta) \tag{5.2}$$

We do not distinguish between sine and cosine in sinusoidal circuit analysis. For brevity, in this book we use cosine functions to describe sinusoidal voltages and currents. If a sinusoid is expressed by a sine function, it is converted to the corresponding cosine function as

$$\sin(\omega t + \theta) = \cos(\omega t + \theta - 90°) \tag{5.3}$$

In AC circuit analysis we often use the term effective value, or the root-mean-square value, to describe voltage and current. They are defined as

$$V_{rms} = \sqrt{\frac{1}{T} \int_0^T v^2(t) dt}$$

$$I_{rms} = \sqrt{\frac{1}{T} \int_0^T i^2(t) dt}$$

For sinusoidal voltages and currents, trigonometric calculation yields

$$V_{rms} = \frac{V_m}{\sqrt{2}} \tag{5.4}$$

$$I_{rms} = \frac{I_m}{\sqrt{2}} \tag{5.5}$$

5.2   Phasor

5.2.1   Complex numbers

In order to simplify the analysis of a sinusoidal circuit, we use phasors, which is a much simpler expression, instead of complex cosine functions to describe sinusoidal voltages and currents. Before introducing phasors, a review of complex numbers is necessary. Suppose there is a complex

到它的极坐标表示形式 $z = r\angle\theta$，这两种表达形式之间的关系为

$$\begin{cases} x = r\cos\theta \\ y = r\sin\theta \end{cases} \qquad \begin{cases} r = \sqrt{x^2 + y^2} \\ \theta = \arctan(y/x) \end{cases} \tag{5.6}$$

在对两个复数 $z_1 = x_1 + jy_1 = r_1\angle\theta_1$ 和 $z_2 = x_2 + jy_2 = r_2\angle\theta_2$ 进行加减运算时，采用代数形式比较方便，有

$$z_1 \pm z_2 = (x_1 \pm x_2) + j(y_1 \pm y_2) \tag{5.7}$$

而进行乘除运算时，采用极坐标形式比较方便，有

$$z_1 z_2 = r_1 r_2 \angle(\theta_1 + \theta_2) \tag{5.8}$$

$$\frac{z_1}{z_2} = \frac{r_1}{r_2}\angle(\theta_1 - \theta_2) \tag{5.9}$$

复数的第三种表示形式为指数形式。将式（5.6）代入复数 $z = x + jy$ 的实部和虚部，得到

$$z = r\cos\theta + jr\sin\theta = r(\cos\theta + j\sin\theta) \tag{5.10}$$

根据欧拉公式

$$e^{\pm j\theta} = \cos\theta \pm j\sin\theta \tag{5.11}$$

复数 z 又可表示为

$$z = re^{j\theta} \tag{5.12}$$

即为复数的指数形式。

5.2.2　相量法

正弦电压为

$$v_1(t) = V_1 \cos(\omega t + \theta_1)$$

其相量定义为

$$V_1 = V_1\angle\theta_1$$

式中，V_1 为电压幅值；θ_1 为电压的相角。

通常用黑斜体字母表示相量。在正弦电路中，如果各电压源和电流源均提供相同频率的正弦量，则各支路的电压和电流也是同一频率的正弦量，因此在相量法中没有出现频率 ω。若电压以正弦函数的形式出现，即

$$v_2(t) = V_2 \sin(\omega t + \theta_2)$$

那么在应用相量法之前需要将其转化为余弦函数形式，即

$$v_2(t) = V_2 \cos(\omega t + \theta_2 - 90°)$$

其相量为

$$V_2 = V_2\angle(\theta_2 - 90°)$$

正弦电流的相量表示也遵循同样的原则。电流

$$i_1(t) = I_1 \cos(\omega t + \varphi_1)$$

number $z = x + jy$, where x is the real part, y denotes the imaginary part, and $j = \sqrt{-1}$ is the imaginary unit (in mathematics, the imaginary unit is denoted by i; however, in circuit analysis, it is denoted by j as i is used to indicate current). $z = x + jy$ is the algebraic form of the complex number. If we draw z in a complex plane, as shown in Fig. 5.2, we can obtain its polar form $z = r\angle\theta$. The relationship between the two forms of expression is

$$\begin{cases} x = r\cos\theta \\ y = r\sin\theta \end{cases} \quad \begin{cases} r = \sqrt{x^2 + y^2} \\ \theta = \arctan(y/x) \end{cases} \tag{5.6}$$

For the addition and subtraction of two complex numbers $z_1 = x_1 + jy_1 = r_1\angle\theta_1$ and $z_2 = x_2 + jy_2 = r_2\angle\theta_2$, we prefer to use their algebraic form

$$z_1 \pm z_2 = (x_1 \pm x_2) + j(y_1 \pm y_2) \tag{5.7}$$

To compute their multiplication and division, a wiser choice is to use their polar form

$$z_1 z_2 = r_1 r_2 \angle(\theta_1 + \theta_2) \tag{5.8}$$

$$\frac{z_1}{z_2} = \frac{r_1}{r_2} \angle(\theta_1 - \theta_2) \tag{5.9}$$

A complex number can also be expressed in exponential form. For a complex number $z = x + jy$, substituting its real and imaginary parts by (5.6), we have

$$z = r\cos\theta + jr\sin\theta = r(\cos\theta + j\sin\theta) \tag{5.10}$$

According to Euler's identity

$$e^{\pm j\theta} = \cos\theta \pm j\sin\theta \tag{5.11}$$

complex number z can be written as

$$z = re^{j\theta} \tag{5.12}$$

which is the so-called exponential form.

5.2.2　Phasors

The phasor of a sinusoidal voltage

$$v_1(t) = V_1 \cos(\omega t + \theta_1)$$

is defined as

$$\boldsymbol{V}_1 = V_1 \angle\theta_1$$

where V_1 represents the amplitude of the voltage and θ_1 represents its phase angle. Usually we use boldface letters for phasors. In a sinusoidal circuit, if all the sources supply voltages or currents of the same frequency, the voltages and currents of the branches also share the same frequency. There fore, angular frequency ω does not appear in phasors. If a voltage is expressed by a sine function

$$v_2(t) = V_2 \sin(\omega t + \theta_2)$$

it should be transformed into the corresponding cosine function before finding its phasor

$$v_2(t) = V_2 \cos(\omega t + \theta_2 - 90°)$$

and its phasor is

$$\boldsymbol{V}_2 = V_2 \angle(\theta_2 - 90°)$$

Phasors for sinusoidal currents follow the same principles. For currents

$$i_1(t) = I_1 \cos(\omega t + \varphi_1)$$

和

$$i_2(t) = I_2 \sin(\omega t + \varphi_2)$$

的相量分别为

$$\boldsymbol{I}_1 = I_1 \angle \varphi_1$$

和

$$\boldsymbol{I}_2 = I_2 \angle (\varphi_2 - 90°)$$

可见，正弦电压和电流的相量形式与复数的极坐标形式是一样的，这就方便我们对正弦量进行计算。假设有如图 5.3 所示的三个同频电压源串联，根据 KVL，其端口电压为

$$v_1(t) = 5\cos(\omega t)\text{V}$$
$$v_2(t) = 10\sin(\omega t + 30°)\text{V}$$
$$v(t)$$
$$v_3(t) = 5\cos(\omega t - 90°)\text{V}$$

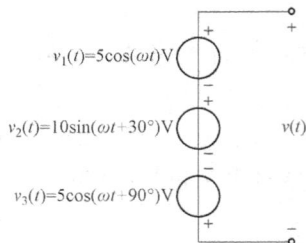

图 5.3　三个同频正弦电压源串联

Fig. 5.3　Three AC voltage sources of the same frequency connected in series

$$v(t) = 5\cos(\omega t) + 10\sin(\omega t + 30°) - 5\cos(\omega t + 90°)(\text{V})$$
$$= 5\cos(\omega t) + 10\cos(\omega t - 60°) - 5\cos(\omega t + 90°)(\text{V}) \tag{5.13}$$

根据欧拉公式可知，$\cos\theta$ 为复数 $\mathrm{e}^{\pm\mathrm{j}\theta}$ 的实部，即

$$\cos\theta = \mathrm{Re}(\mathrm{e}^{\pm\mathrm{j}\theta}) \tag{5.14}$$

则式（5.13）可改写为

$$v(t) = 5\,\mathrm{Re}(\mathrm{e}^{\mathrm{j}\omega t}) + 10\,\mathrm{Re}[\mathrm{e}^{\mathrm{j}(\omega t - 60°)}] - 5\,\mathrm{Re}[\mathrm{e}^{\mathrm{j}(\omega t + 90°)}]$$
$$= \mathrm{Re}(5\mathrm{e}^{\mathrm{j}\omega t}) + \mathrm{Re}[10\mathrm{e}^{\mathrm{j}(\omega t - 60°)}] - \mathrm{Re}[5\mathrm{e}^{\mathrm{j}(\omega t + 90°)}]$$
$$= \mathrm{Re}(5\mathrm{e}^{\mathrm{j}\omega t} + 10\mathrm{e}^{\mathrm{j}\omega t}\mathrm{e}^{-\mathrm{j}60°} - 5\mathrm{e}^{\mathrm{j}\omega t}\mathrm{e}^{\mathrm{j}90°})$$
$$= \mathrm{Re}[(5 + 10\mathrm{e}^{-\mathrm{j}60°} - 5\mathrm{e}^{\mathrm{j}90°})\mathrm{e}^{\mathrm{j}\omega t}]\,\text{V}$$

将上式中复数的指数形式以极坐标形式代替，得到

$$v(t) = \mathrm{Re}[(5 + 10\angle - 60° - 5\angle 90°)\mathrm{e}^{\mathrm{j}\omega t}]\,\text{V} \tag{5.15}$$

将上式中的复数部分转化为代数形式进行计算

$$5 + 10\angle - 60° - 5\angle 90° = 5 + 10\cos(-60°) + \mathrm{j}10\sin(-60°) - 5\cos 90° - \mathrm{j}5\sin 90°$$
$$= 5 + 5 - \mathrm{j}8.66 - \mathrm{j}5 = 10 - \mathrm{j}13.66$$
$$= 16.93\angle - 53.79° = 16.93\mathrm{e}^{-\mathrm{j}53.79°}\,\text{V}$$

则式（5.15）变为

$$v(t) = \mathrm{Re}[16.93\mathrm{e}^{-\mathrm{j}53.79°}\mathrm{e}^{\mathrm{j}\omega t}] = \mathrm{Re}[16.93\mathrm{e}^{\mathrm{j}(-53.79° + \omega t)}]\,\text{V}$$

由欧拉公式可知，上式可写为如下的余弦函数形式

$$v(t) = 16.93\cos(\omega t - 53.79°)\,\text{V}$$

根据上面例子可以总结出利用相量法对同频正弦量进行计算的步骤为：

（1）确定各正弦量的相量：$V_1 = 5\angle 0°\text{V}$，$V_2 = 10\angle - 60°\text{V}$，$V_3 = 5\angle 90°\text{V}$；

（2）将这些相量看作复数，并利用复数的代数形式进行计算

$$V = V_1 + V_2 - V_3 = 5 + 5 - \mathrm{j}8.66 - \mathrm{j}5 = 10 - \mathrm{j}13.66(\text{V})$$

（3）将上面计算结果的代数形式转换为极坐标形式为

$$V = 10 - \mathrm{j}13.66 = 16.93\angle - 53.79°\,(\text{V})$$

and

$$i_2(t) = I_2 \sin(\omega t + \varphi_2)$$

the phasors are

$$\boldsymbol{I}_1 = I_1 \angle \varphi_1$$

and

$$\boldsymbol{I}_2 = I_2 \angle (\varphi_2 - 90°)$$

Apparently, phasors for sinusoidal voltages and currents have the same format as the polar form of a complex number, which makes the calculation of sinusoids more convenient. Suppose we have three sinusoidal voltage sources connected in series, as shown in Fig. 5.3, and they have the same frequency. According to KVL, the output voltage is

$$v(t) = 5\cos(\omega t) + 10\sin(\omega t + 30°) - 5\cos(\omega t + 90°)$$

$$= 5\cos(\omega t) + 10\cos(\omega t - 60°) - 5\cos(\omega t + 90°)(V)$$

(5.13)

Euler's identity (5.11) indicates that $\cos\theta$ is the real part of the complex number $e^{\pm j\theta}$:

$$\cos\theta = \text{Re}(e^{\pm j\theta})$$

(5.14)

Therefore, (5.13) can be written as

$$v(t) = 5\,\text{Re}(e^{j\omega t}) + 10\,\text{Re}[e^{j(\omega t - 60°)}] - 5\,\text{Re}[e^{j(\omega t + 90°)}]$$

$$= \text{Re}(5e^{j\omega t}) + \text{Re}[10e^{j(\omega t - 60°)}] - \text{Re}[5e^{j(\omega t + 90°)}]$$

$$= \text{Re}(5e^{j\omega t} + 10e^{j\omega t}e^{-j60°} - 5e^{j\omega t}e^{j90°})$$

$$= \text{Re}[(5 + 10e^{-j60°} - 5e^{j90°})e^{j\omega t}]\,V$$

Using the polar form instead of the exponential form to represent the complex numbers, we have

$$v(t) = \text{Re}[(5 + 10\angle - 60° - 5\angle 90°)e^{j\omega t}]\,V$$

(5.15)

To combine the complex numbers, we can use their algebraic form

$$5 + 10\angle - 60° - 5\angle 90° = 5 + 10\cos(-60°) + j10\sin(-60°) - 5\cos 90° - j5\sin 90°$$

$$= 5 + 5 - j8.66 - j5 = 10 - j13.66$$

$$= 16.93\angle - 53.79° = 16.93e^{-j53.79°}\,V$$

And (5.15) can be written as

$$v(t) = \text{Re}[16.93e^{-j53.79°}e^{j\omega t}] = \text{Re}[16.93e^{j\,(-53.79° + \omega t)}]\,V$$

Again we apply Euler's identity, the voltage can be written as an expression of time

$$v(t) = 16.93\cos(\omega t - 53.79°)\,V$$

From the above example, we can find a way for calculating sinusoids with the same frequency, which is

(1) Determine the phasor for each sinusoid: $V_1 = 5\angle 0°\,V$, $V_2 = 10\angle - 60°\,V$, $V_3 = 5\angle 90°\,V$;

(2) View these phasors as complex numbers and perform calculation using their algebraic form

$$V = V_1 + V_2 - V_3 = 5 + 5 - j8.66 - j5 = 10 - j13.66(V)$$

(3) Transform the result into the polar form

$$V = 10 - j13.66 = 16.93\angle - 53.79°(V)$$

（4）将上述以极坐标形式描述的复数看作相量，并据此写出时域方程

$$v(t) = 16.93\cos(\omega t - 53.79°) \text{ V}$$

5.3 阻抗与导纳

5.3.1 电阻、电感和电容的电压—电流相量的关系

在用相量法对电压和电流进行描述后，接下来讨论正弦稳态电路中各负载元件（电阻、电感、电容）的以相量表示的电压和电流之间的关系。首先讨论电阻。假设有电流 $i(t) = I\cos(\omega t + \theta)$ 流过电阻 R，则根据欧姆定律，该电阻两端电压为

$$v(t) = Ri(t) = RI\cos(\omega t + \theta)$$

电流与电压的相量分别为 $\boldsymbol{I} = I\angle\theta$ 和 $\boldsymbol{V} = RI\angle\theta$，显然有

$$\boldsymbol{V} = R\boldsymbol{I} \tag{5.16}$$

可见对于电阻而言，其电压与电流同相位。

假设有电流 $i(t) = I\cos(\omega t + \theta)$ 流过电感 L，则根据式（4-18），电感两端电压为

$$v(t) = L\frac{\mathrm{d}i}{\mathrm{d}t} = -LI\omega\sin(\omega t + \theta) = LI\omega\cos(\omega t + \theta + 90°)$$

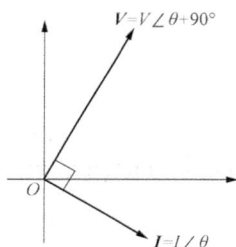

图 5.4　电感的电压与电流的相量图

Fig. 5.4　The phasor diagram for voltage and current of an inductor

其相量为

$$\boldsymbol{V} = LI\omega\angle(\theta + 90°) = V\angle(\theta + 90°) \tag{5.17}$$

可见电感电压的相位超前电流相位 90°，如图 5.4 所示。

对式（5.17）做如下变换

$$\boldsymbol{V} = LI\omega\angle(\theta + 90°) = \omega L\angle 90° \cdot I\angle\theta = \mathrm{j}\omega LI$$

可得相量法表示下电感的电压与电流关系

$$\boldsymbol{V} = \mathrm{j}\omega L\boldsymbol{I} \tag{5.18}$$

式中，$\mathrm{j}\omega L$ 为电感的阻抗，其实部为零，仅含有虚部，且虚部为正数。

对于电容 C，我们假设其两端电压为 $v(t) = V\cos(\omega t + \theta)$，则根据式（4.3），流过电容的电流为

$$i(t) = C\frac{\mathrm{d}v}{\mathrm{d}t} = -CV\omega\sin(\omega t + \theta) = CV\omega\cos(\omega t + \theta + 90°)$$

分别用相量表示电压和电流

$$\boldsymbol{V} = V\angle\theta \tag{5.19}$$

$$\boldsymbol{I} = CV\omega\angle(\theta + 90°) = I\angle(\theta + 90°) \tag{5.20}$$

可见电容电流的相位超前电压 90°，如图 5.5 所示。

对式（5.20）做如下改写

$$\boldsymbol{I} = C\omega\angle 90° \cdot V\angle\theta = \mathrm{j}\omega C\boldsymbol{V} \tag{5.21}$$

(4) The complex number in its polar form is considered as the phasor of the result, based on which we can rewrite above as a function of time

$$v(t) = 16.93\cos(\omega t - 53.79°) \text{ V}$$

5.3　Impedance and Admittance

5.3.1　Voltage-current relationship in phasor for resistors, inductors and capacitors

After describing voltages and currents in phasors, we continue to discuss the voltage-current relationship, in phasors, of passive circuit elements (resistors, capacitors and inductors) for steady-state sinusoidal circuits. Let's start from resistors. Suppose current $i(t) = I\cos(\omega t + \theta)$ flows through resistor R. According to Ohm's law, its voltage should be

$$v(t) = Ri(t) = RI\cos(\omega t + \theta)$$

The phasors of the current and the voltage are $\boldsymbol{I} = I\angle\theta$ and $\boldsymbol{V} = RI\angle\theta$, respectively, which gives

$$\boldsymbol{V} = R\boldsymbol{I} \tag{5.16}$$

For a resistor, its voltage and current are of the same phase.

Suppose current $i(t) = I\cos(\omega t + \theta)$ flows through an inductor L. Referring to (4.18), the voltage across the inductor is

$$v(t) = L\frac{di}{dt} = -LI\omega\sin(\omega t + \theta) = LI\omega\cos(\omega t + \theta + 90°)$$

and its phasor is

$$\boldsymbol{V} = LI\omega\angle(\theta + 90°) = V\angle(\theta + 90°) \tag{5.17}$$

For an inductor, the phase of its voltage leads the phase of its current by 90°, as illustrated in Fig. 5.4.

Equation (5.17) can be written in the following way

$$\boldsymbol{V} = LI\omega\angle(\theta + 90°) = \omega L\angle 90° \cdot I\angle\theta = j\omega L\cdot\boldsymbol{I}$$

From the above equation we can draw the voltage-current relationship in phasors of an inductor as

$$\boldsymbol{V} = j\omega L\boldsymbol{I} \tag{5.18}$$

where $j\omega L$ is called the impedance of the inductor; its real part is zero and its imaginary part is positive.

For a capacitor C, we assume a voltage $v(t) = V\cos(\omega t + \theta)$ is applied across it. According to (4.3), the current flowing through it is

$$i(t) = C\frac{dv}{dt} = -CV\omega\sin(\omega t + \theta) = CV\omega\cos(\omega t + \theta + 90°)$$

Expressing the voltage and the current in phasors, we have

$$\boldsymbol{V} = V\angle\theta \tag{5.19}$$
$$\boldsymbol{I} = CV\omega\angle(\theta + 90°) = I\angle(\theta + 90°) \tag{5.20}$$

It can be seen that for a capacitor, the phase of its current leads the phase of its voltage by 90°, as shown in Fig. 5.5.

Writing (5.20) in the following manner

$$\boldsymbol{I} = C\omega\angle 90° \cdot V\angle\theta = j\omega C\boldsymbol{V} \tag{5.21}$$

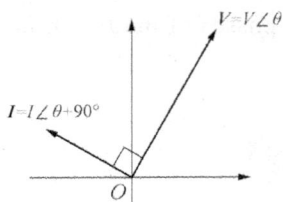

图 5.5　电容的电压与电流的相量图

Fig. 5.5　The phasor diagram for voltage and current of a capacitor

可得用相量表示的电容电压与电流关系

$$V = \frac{I}{j\omega C} \tag{5.22}$$

式中，$\dfrac{1}{j\omega C}$ 为电容的阻抗，其实部为零，仅含有虚部，且虚部为负数。

5.3.2　阻抗

上节中我们得到了电阻、电感和电容的电压—电流关系分别为

$$V = RI \quad V = j\omega LI \quad V = \frac{I}{j\omega C} = -\frac{j}{\omega C}I$$

可见任何元件的电压—电流关系都可通过相量形式的欧姆定律获得

$$V = ZI \tag{5.23}$$

式中，Z 为该元件的阻抗，单位为欧（Ω）。

当电路负载为三种基本元件的组合时，其阻抗具有如下形式

$$Z = R + jX \tag{5.24}$$

式中，实部 R 为电阻；虚部 X 为电抗。电阻 R 一定满足 $R \geqslant 0$，当 $R = 0$ 时，表示负载为纯电感或纯电容；电抗 X 可为正数、负数或零，$X > 0$ 说明负载是感性的，$X < 0$ 说明负载是容性的，$X = 0$ 则说明负载为纯电阻。

阻抗也可表示为极坐标的形式

$$Z = |Z| \angle \theta \tag{5.25}$$

式中，$|Z| = \sqrt{R^2 + X^2}$ 为阻抗模；$\theta = \arctan \dfrac{X}{R}$，反映了电压与电流的相位差，且有 $R = |Z| \cos\theta$，$X = |Z| \sin\theta$。

5.3.3　导纳

导纳用 Y 表示，其定义为阻抗的倒数，即

$$Y = \frac{1}{Z} = \frac{I}{V} \tag{5.26}$$

导纳的单位为 S（西门子）。导纳同样为复数，因此可表示为

$$Y = G + jB \tag{5.27}$$

式中，实部 G 称为电导；虚部 B 称为电纳。

根据

$$Y = G + jB = \frac{1}{Z} = \frac{1}{R + jX} = \frac{R - jX}{R^2 + X^2}$$

可得

$$G = \frac{R}{R^2 + X^2}$$

$$B = -\frac{X}{R^2 + X^2}$$

we can find that their phasors are related as

$$V = \frac{I}{j\omega C} \tag{5.22}$$

where $\dfrac{1}{j\omega C}$ is the impedance of the capacitor, which also has a zero real part but a negative imaginary part.

5.3.2 Impedance

The voltage-current relationship of resistors, inductors and capacitors, respectively, are listed below

$$V = RI \quad V = j\omega LI \quad V = \frac{I}{j\omega C} = -\frac{j}{\omega C} I$$

Examining these equations we can conclude that for any passive element, its voltage-current relationship can be expressed by the Ohm's law in phasors

$$V = ZI \tag{5.23}$$

where Z is the impedance of this element with the unit of ohms (Ω). If the load of a circuit is the combination of resistors, inductors and capacitors, its impedance has the following expression

$$Z = R + jX \tag{5.24}$$

where the real part, R, represents the resistance and the imaginary part, X, stands for the reactance.

The value of the resistance must satisfy $R \geqslant 0$; when $R = 0$, the load is pure inductance or pure capacitance. The reactance can have any value, positive, zero or negative; $X > 0$ indicates the load is inductive, $X < 0$ indicates it is capacitive, and $X = 0$ means the load is pure resistance.

The impedance can also be expressed in the polar form

$$Z = |Z| \angle \theta \tag{5.25}$$

where $|Z| = \sqrt{R^2 + X^2}$ is the magnitude, $\theta = \arctan \dfrac{X}{R}$ gives the phase difference between voltage and current, $R = |Z| \cos\theta$, and $X = |Z| \sin\theta$.

5.3.3 Admittance

Admittance, denoted by Y, is defined as the reciprocal of the impedance

$$Y = \frac{1}{Z} = \frac{I}{V} \tag{5.26}$$

It has the unit of siemens (S). The admittance is also a complex number, which can be expressed by

$$Y = G + jB \tag{5.27}$$

where the real part G is the conductance and the imaginary part B is the susceptance. According to the following equation

$$Y = G + jB = \frac{1}{Z} = \frac{1}{R + jX} = \frac{R - jX}{R^2 + X^2}$$

we have

$$G = \frac{R}{R^2 + X^2}$$

$$B = -\frac{X}{R^2 + X^2}$$

显然对于非纯电阻负载，$G = 1/R$ 不成立。

5.4　正弦电路的稳态分析

正弦稳态电路的分析方法与在本书前 3 章中所学的电阻电路分析方法并无本质区别，学过的所有电路分析原理和方法均可应用于正弦稳态电路分析，差别仅在于需要用相量表示电压和电流以及用阻抗描述负载。

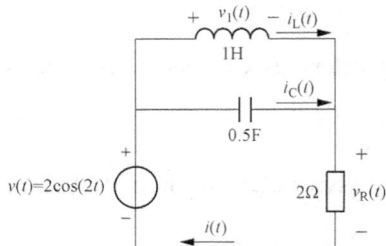

图 5.6　例 5.1 电路图

Fig. 5.6　Circuit diagram for Example 5.1

【例 5.1】　考虑如图 5.6 所示电路，电源电压为 $v(t) = 2\cos(2t)$ V，各负载元件参数分别为 $L=1$ H、$C=0.5$ F、$R = 2$ Ω。求电流 $i_C(t)$、$i_L(t)$ 和 $i(t)$ 的相量。

解　电压源输出电压的相量为 $V = 2\angle 0°$ V，角频率为 $\omega = 2$ rad，电感、电容和电阻三个元件的阻抗分别为

$$Z_L = j\omega L = j2 = 2\angle 90°\ \Omega$$

$$Z_C = \frac{1}{j\omega C} = -j = 1\angle -90°\ \Omega$$

$$Z_R = 2\ \Omega$$

电感与电容并联的等效阻抗为

$$Z_1 = \frac{Z_L Z_C}{Z_L + Z_C} = \frac{j2 \cdot (-j)}{j2 - j} = -j2 = 2\angle -90°\ \Omega$$

其求解公式与电阻并联相同。根据分压原理，电容（电感）两端电压为

$$V_1 = \frac{Z_1}{Z_1 + Z_R} V = \frac{-j2}{-j2 + 2} \cdot 2\angle 0° = \frac{2\angle -90°}{2\sqrt{2}\angle -45°} \cdot 2\angle 0° = \sqrt{2}\angle -45°\ \text{V}$$

根据欧姆定律，电容和电感的电流分别为

$$I_C = \frac{V_1}{Z_C} = \frac{\sqrt{2}\angle -45°}{1\angle -90°} = \sqrt{2}\angle 45°\ \text{A}$$

$$I_L = \frac{V_1}{Z_L} = \frac{\sqrt{2}\angle -45°}{2\angle 90°} = \frac{\sqrt{2}}{2}\angle -135°\ \text{A}$$

根据 KCL，流过电阻的电流为

$$I = I_C + I_L = \sqrt{2}\angle 45° + \frac{\sqrt{2}}{2}\angle -135°$$

$$= 1 + j - 0.5 - 0.5j = 0.5 + j0.5 = \frac{\sqrt{2}}{2}\angle 45°\ \text{A}$$

也可通过欧姆定律计算该电流

$$I = \frac{V}{Z_1 + Z_R} = \frac{2\angle 0°}{-j2 + 2} = \frac{2\angle 0°}{2\sqrt{2}\angle -45°} = \frac{\sqrt{2}}{2}\angle 45°\ \text{A}$$

通过两种方法求得的结果一致。

It should be noted that the relationship $G=1/R$ only holds for pure resistors.

5.4　Steady–state Sinusoidal Analysis

The methods we use to solve for steady-state sinusoidal circuits are basically the same as the ones we used for DC circuits. The laws, principals and circuit analysis methods we have learnt in the first three chapters can also be applied in solving steady-state sinusoidal circuits. The only difference lies in that we use phasors to describe voltages and currents and impedances to describe loads.

Example 5.1: Consider the circuit shown in Fig. 5.6, in which the source voltage is $v(t) = 2\cos(2t)$ and the values of the loads are $L = 1\,\text{H}$, $C = 0.5\,\text{F}$ and $R = 2\,\Omega$, respectively. Find the phasors for $i_C(t)$, $i_L(t)$ and $i(t)$.

Solution:

The phasor for the source voltage is $V = 2\text{V}$, the angular frequency is $\omega = 2$ rad, and the impedances of the inductor, capacitor and resistor, respectively, are

$$Z_L = j\omega L = j2 = 2\angle 90°\,\Omega$$

$$Z_C = \frac{1}{j\omega C} = -j = 1\angle -90°\,\Omega$$

$$Z_R = 2\,\Omega$$

In order to find out the equivalent impedance of the inductance and the capacitance connected in parallel, we apply the same method as we calculate the equivalent resistance of parallel connected resistors

$$Z_1 = \frac{Z_L Z_C}{Z_L + Z_C} = \frac{j2 \cdot (-j)}{j2 - j} = -j2 = 2\angle -90°\,\Omega$$

According to the voltage division principle, the voltage across the capacitor (inductor) is

$$V_1 = \frac{Z_1}{Z_1 + Z_R}V = \frac{-j2}{-j2 + 2} \cdot 2\angle 0° = \frac{2\angle -90°}{2\sqrt{2}\angle -45°} \cdot 2\angle 0° = \sqrt{2}\angle -45°\text{V}$$

Therefore, according to Ohm's law, the phasor currents of the capacitor and the inductor, respectively, are

$$I_C = \frac{V_1}{Z_C} = \frac{\sqrt{2}\angle -45°}{1\angle -90°} = \sqrt{2}\angle 45°\text{A}$$

$$I_L = \frac{V_1}{Z_L} = \frac{\sqrt{2}\angle -45°}{2\angle 90°} = \frac{\sqrt{2}}{2}\angle -135°\text{A}$$

Consequently, the phasor current of the resistor can be obtained by applying KCL

$$I = I_C + I_L = \sqrt{2}\angle 45° + \frac{\sqrt{2}}{2}\angle -135°$$

$$= 1 + j - 0.5 - 0.5j = 0.5 + j0.5 = \frac{\sqrt{2}}{2}\angle 45°\text{A}$$

We can also calculate the current by applying Ohm's law

$$I = \frac{V}{Z_1 + Z_R} = \frac{2\angle 0°}{-j2 + 2} = \frac{2\angle 0°}{2\sqrt{2}\angle -45°} = \frac{\sqrt{2}}{2}\angle 45°\text{A}$$

We achieve the same result using the two methods.

【例 5.2】 戴维南和诺顿等效定理是电路分析的重要方法，它们在正弦稳态电路分析中具有与直流电路分析中相同的形式。以图 5.7（a）所示电路为例，求其戴维南和诺顿等效电路。

解 首先将电路中的独立电源置零以求戴维南阻抗，如图 5.7（b）所示，戴维南阻抗 Z_t 为电容与电阻的并联

$$Z_t = \frac{Z_R Z_C}{Z_R + Z_C} = \frac{100 \cdot (-j100)}{100 - j100} = \frac{100\angle -90°}{\sqrt{2}\angle -45°}$$
$$= 50\sqrt{2}\angle -45° = 50 - j50\ \Omega$$

接下来，将输出端口短路以求短路电流 I_{sc}，如图 5.7（c）所示。由于电容被短路，其两端电压为零，从而 $I_c = 0$。根据 KCL 可列出如下方程

$$I_{sc} = I_R - I_s = \frac{V_s}{Z_R} - I_s = \frac{100\angle 0°}{100} - 1\angle 90° = 1 - j = \sqrt{2}\angle -45°\ \text{A}$$

此短路电流即为诺顿等效电路的电源电流 I_n。进而可得戴维南电压为

$$V_t = I_{sc} Z_t = \sqrt{2}\angle -45° \cdot 50\sqrt{2}\angle -45° = 100\angle -90°\ \text{V}$$

请读者根据图 5.7（e）计算开路电压 V_{oc}，V_{oc} 是否等于 V_t？

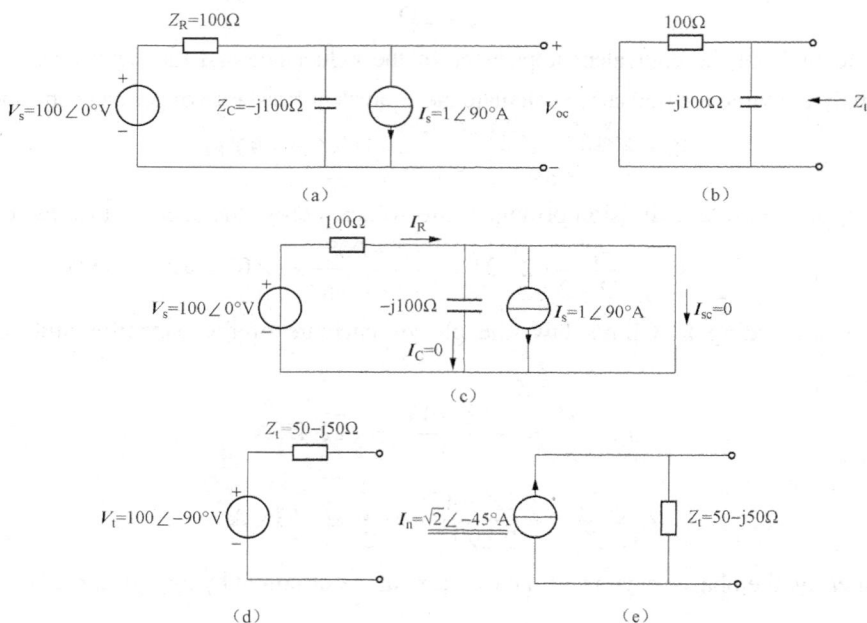

图 5.7 例 5.2 电路图

（a）原始电路；（b）求戴维南阻抗的电路；

（c）求短路电流的电路；（d）戴维南等效电路；（e）诺顿等效电路

Fig. 5.7 Circuit diagrams for Example 5.2

(a) Original circuit; (b) Calculating Théve nin impedance; (c) Circuit for finding short circuit current;

(d) Thévenin equivalent circuit; (e) Norton equivalent circuit

Example 5.2: Thévenin and Norton equivalent circuits are useful ways for circuit analysis. They are applied in the same manner in steady-state sinusoidal circuits as in DC circuits. Find the Thévenin and Norton equivalents for the circuit shown in Fig. 5.7(a).

Solution:

We zero all the independent sources to find out the Thévenin impedance. As shown in Fig. 5.7(b), the Thévenin impedance Z_t is the parallel combination of the capacitor and the resistor

$$Z_t = \frac{Z_R Z_C}{Z_R + Z_C} = \frac{100 \cdot (-j100)}{100 - j100} = \frac{100\angle -90°}{\sqrt{2}\angle -45°} = 50\sqrt{2}\angle -45° = 50 - j50 \ \Omega$$

Next, we connect the two terminals by a short ciruit to solve for the short circuit current I_{sc}, as depicted in Fig. 5.7(c). As the capacitor is shorted, its voltage is zero; thus $I_c = 0$. According to KCL, the following equation can be written

$$I_{sc} = I_R - I_s = \frac{V_s}{Z_R} - I_s = \frac{100\angle 0°}{100} - 1\angle 90° = 1 - j1 = \sqrt{2}\angle -45° \ \text{A}$$

This short circuit current is the source current I_n in the Norton equivalent. We can therefore obtain the Thévenin voltage by

$$V_t = I_{sc} Z_t = \sqrt{2}\angle -45° \cdot 50\sqrt{2}\angle -45° = 100\angle -90° \ \text{V}$$

We can also solve for the open circuit voltage V_{oc} from the circuit shown in Fig. 5.7(e) and check whether V_{oc} equals to V_t. We leave this to the readers to finish.

Charles Proteus Steinmetz (9 April 1865 – 26 October 1923) was a German-born American mathematician and electrical engineer. He fostered the development of alternating current that made the expansion of the electric power industry in the United States possible, and formulated mathematical theories for engineers. He also founded the phasor method, which is a practical method for calculating the AC circuit. Before Steinmetz's groundbreaking work, engineers had been using complicated, time-consuming calculus-based methods. Steinmetz systematized the use of complex number phasor representation, which revolutionized AC circuit theory and analysis.

5.5　交流电路功率

5.5.1　瞬时功率

功率计算也是交流电路分析的一项重要内容。如第 1 章中定义的，一个电路元件在某时刻吸收的功率（即瞬时功率）为此时刻元件两端电压与流过它的电流的乘积。在电压和电流取关联参考方向的情况下，瞬时功率由下式计算

$$p(t) = v(t)i(t) \tag{5.28}$$

假设瞬时电压和电流的表达式为

$$v(t) = V_{\mathrm{m}} \cos(\omega t + \theta_v)$$
$$i(t) = I_{\mathrm{m}} \cos(\omega t + \theta_i)$$

将其代入式（5.28）中，得到

$$p(t) = v(t)i(t) = V_{\mathrm{m}} I_{\mathrm{m}} \cos(\omega t + \theta_v) \cos(\omega t + \theta_i)$$

根据三角函数积化和差公式，上式变为

$$p(t) = \frac{1}{2} V_{\mathrm{m}} I_{\mathrm{m}} \cos(\theta_v - \theta_i) + \frac{1}{2} V_{\mathrm{m}} I_{\mathrm{m}} \cos(2\omega t + \theta_v + \theta_i) \tag{5.29}$$

上式右边第一项为常数，表现在波形上是导致整条功率曲线向上偏移，如图 5.8 所示。式（5.29）右边第二项是一个正弦函数，其角频率为 2ω，是电压和电流角频率的两倍。显然瞬时功率的周期为电压和电流周期的一半。

对于电阻，$\theta_v = \theta_i$，因此 $\cos(\theta_v - \theta_i) = 1$，如图 5.9 所示，其瞬时功率 $p(t) \geqslant 0$。而对于电容和电感，$\theta_v - \theta_i = \mp 90°$，$\cos(\theta_v - \theta_i) = 0$，其瞬时功率半周期为正半周期为负，如图 5.10 所示。

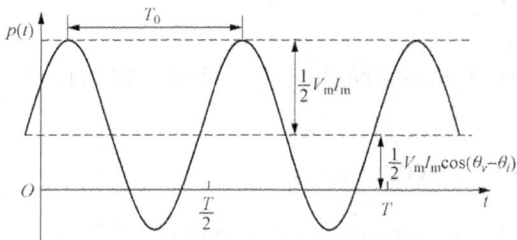

图 5.8　瞬时功率波形示意图

Fig. 5.8　A plot of instantaneous power

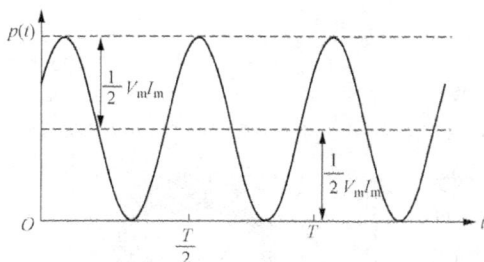

图 5.9　电阻瞬时功率波形示意图

Fig. 5.9　A plot of instantaneous power for a resistor

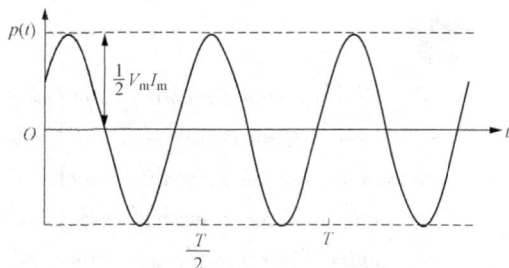

图 5.10　电容和电感的瞬时功率波形示意图

Fig. 5.10　A plot of instantaneous power for a capacitor or an inductor

5.5.2　平均功率

平均功率定义为瞬时功率在一个周期内的平均值。这里的周期可以是电压和电流的周期 T，也可以是功率的周期 $T_0 = T/2$，其公式为

$$P = \frac{1}{T} \int_0^T p(t)\mathrm{d}t \tag{5.30}$$

5.5 Power in AC Circuits

5.5.1 Instantaneous power

Power analysis is of great importance in sinusoidal circuit analysis. As defined in Chapter 1, power absorbed by a circuit element at a time instant (instantaneous power) equals to the product of the voltage across it and the current flows through it at that very instant in time. When the voltage and the current have passive reference configuration, the instantaneous power can be calculated from

$$p(t) = v(t)i(t) \tag{5.28}$$

Suppose the instantaneous voltage and current are given as

$$v(t) = V_{\text{m}} \cos(\omega t + \theta_v)$$

$$i(t) = I_{\text{m}} \cos(\omega t + \theta_i)$$

Substituting them into (5.28), we have

$$p(t) = v(t)i(t) = V_{\text{m}} I_{\text{m}} \cos(\omega t + \theta_v) \cos(\omega t + \theta_i)$$

Applying the trigonometric identity to the right-hand side of the above equation, we obtain

$$p(t) = \frac{1}{2} V_{\text{m}} I_{\text{m}} \cos(\theta_v - \theta_i) + \frac{1}{2} V_{\text{m}} I_{\text{m}} \cos(2\omega t + \theta_v + \theta_i) \tag{5.29}$$

The first term on the right-hand side of (5.29) is constant, showing an upward offset in the waveform plotted in Fig. 5.8. The second term on the right-hand side of (5.29) is a sinusoidal function whose angular frequency is 2ω, which is twice of the angular frequency of voltage and current. Apparently, the period of instantaneous power is half that of voltage and current.

For a resistor, we have $\theta_v = \theta_i$; thus $\cos(\theta_v - \theta_i) = 1$. Therefore, its instantaneous power always satisfies $p(t) \geqslant 0$, as shown in Fig. 5.9. On the other hand, for a capacitor and an inductor, the phase difference between voltage and current is $\theta_v - \theta_i = \mp 90°$; thus $\cos(\theta_v - \theta_i) = 0$. As plotted in Fig. 5.10, the instantaneous power is positive for the first half-cycle and negative for the next half-cycle.

5.5.2 Average power

The average power is defined as the average of the instantaneous power over one period. Here it can either be the period of voltage/current T or the period of instantaneous power $T_0 = T/2$. Thus, the average power is given by

$$P = \frac{1}{T} \int_0^T p(t) \mathrm{d}t \tag{5.30}$$

将式（5.29）的 $p(t)$ 代入上式，有

$$P = \frac{1}{T}\int_0^T \frac{1}{2}V_m I_m \cos(\theta_v - \theta_i)dt + \frac{1}{T}\int_0^T \frac{1}{2}V_m I_m \cos(2\omega t + \theta_v + \theta_i)dt$$

$$= \frac{1}{2}V_m I_m \cos(\theta_v - \theta_i)\frac{1}{T}\int_0^T dt + \frac{1}{2}V_m I_m \frac{1}{T}\int_0^T \cos(2\omega t + \theta_v + \theta_i)dt$$

上式右边第一项中 $\frac{1}{T}\int_0^T dt = 1$，第二项中余弦函数在一个周期内的积分为 0，因此平均功率为

$$P = \frac{1}{2}V_m I_m \cos(\theta_v - \theta_i) \tag{5.31}$$

将正弦电压和电流的有效值表达式（5.4）和式（5.5）代入，平均功率也可写为

$$P = V_{rms} I_{rms} \cos(\theta_v - \theta_i) \tag{5.32}$$

对于电阻而言，其电压与电流同相位，即 $\theta_v = \theta_i$，则由上式可知电阻吸收的平均功率为 $P = \frac{1}{2}V_m I_m$。而对于电容和电感，它们的电压和电流相差 90°，即 $\theta_v - \theta_i = \mp 90°$，因此电容和电感吸收的平均功率为 $P = 0$。平均功率也称为有功功率。

5.5.3　功率因数、无功功率和视在功率

定义式（5.32）中电压与电流的相位差为 $\theta = \theta_v - \theta_i$，则式（5.32）变为

$$P = V_{rms} I_{rms} \cos\theta \tag{5.33}$$

其中 $-90° \leqslant \theta \leqslant 90°$，$0 \leqslant \cos\theta \leqslant 1$。这里 $\cos\theta$ 称为功率因数，即

$$PF = \cos\theta \tag{5.34}$$

通常用百分数表示，并指明是电流超前电压（容性负载）还是电流滞后电压（感性负载）。

如前所述，电容和电感是储能元件，在交流电路中，它们前半周期从电路吸收能量，后半周期将能量返回给电路，如图 5.10 所示。瞬时功率表达式（5.29）可改写为

$$p(t) = \frac{1}{2}V_m I_m \cos(\theta_v - \theta_i)\left[1 + \cos(2\omega t + 2\theta_v)\right] + \frac{1}{2}V_m I_m \sin(\theta_v - \theta_i)\sin(2\omega t + 2\theta_v)$$

上式右边第一项始终大于或等于零，而第二项的值正负交替，说明电路内部有能量在储能元件间进行交换。我们将与能量交换相关的瞬时功率的最大值（即上式右边第二项的最大值）定义为无功功率

$$Q = V_{rms} I_{rms} \sin\theta \tag{5.35}$$

单位为乏（var）。

视在功率定义为电压与电流有效值之积

$$S = V_{rms} I_{rms} \tag{5.36}$$

单位为伏安（VA）。

由式（5.33）、式（5.35）和式（5.36）可知，平均功率、无功功率和视在功率满足如下方程

$$P^2 + Q^2 = S^2 \tag{5.37}$$

构成如图 5.11 所示的功率三角形。

电路中的平均功率和无功功率分别对应着电阻消耗的功率和储能元件（电容和电感）在电路中交换的

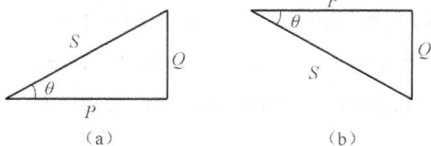

图 5.11　功率三角形

（a）感性负载，电流滞后电压，$\theta > 0$；

（b）容性负载，电流超前电压，$\theta < 0$

Fig. 5.11　Power triangles

(a) Inductive load, current lags voltage, $\theta > 0$;

(b) Capacitive load, current leads voltage, $\theta < 0$

Substituting $p(t)$ in (5.29) into the above equation yields

$$P = \frac{1}{T}\int_0^T \frac{1}{2}V_m I_m \cos(\theta_v - \theta_i)dt + \frac{1}{T}\int_0^T \frac{1}{2}V_m I_m \cos(2\omega t + \theta_v + \theta_i)dt$$

$$= \frac{1}{2}V_m I_m \cos(\theta_v - \theta_i)\frac{1}{T}\int_0^T dt + \frac{1}{2}V_m I_m \frac{1}{T}\int_0^T \cos(2\omega t + \theta_v + \theta_i)dt$$

In the first term on the right-hand side of the above equation, we have $\frac{1}{T}\int_0^T dt = 1$. In the second term, the cosine function integrated over one cycle results in 0. Thus, the average power becomes

$$P = \frac{1}{2}V_m I_m \cos(\theta_v - \theta_i) \tag{5.31}$$

If we use the effective value (5.4) and (5.5) to describe voltage and current, the average power can be written as

$$P = V_{rms}I_{rms}\cos(\theta_v - \theta_i) \tag{5.32}$$

As we know the voltage and current of a resistor are in phase, i.e. $\theta_v = \theta_i$. Hence, the average power absorbed by a resistor is $P = \frac{1}{2}V_m I_m$. On the other hand, for a capacitor or an inductor, there is a 90° difference between its voltage and current, i.e. $\theta_v - \theta_i = \mp 90°$. Therefore, the average power absorbed by capacitor and inductor is $P = 0$.

5.5.3 Power factor, reactive power and apparent power

Define the phase difference in (5.32) by $\theta = \theta_v - \theta_i$, then (5.32) becomes

$$P = V_{rms}I_{rms}\cos\theta \tag{5.33}$$

where $-90° \leqslant \theta \leqslant 90°$ and $0 \leqslant \cos\theta \leqslant 1$. The term $\cos\theta$ is called the power factor

$$PF = \cos\theta \tag{5.34}$$

Power factor is usually expressed in percentage and it is common to state whether the current leads the voltage (capacitive load) or the current lags the voltage (inductive load).

As previously explained, capacitors and inductors are energy-storage elements. In an AC circuit, they absorb energy from the circuit in the first half-cycle and return the energy to the circuit in the second half-cycle, as shown in Fig. 5.10. Instantaneous power(5.9) can be rewritten as

$$p(t) = \frac{1}{2}V_m I_m \cos(\theta_v - \theta_i)\left[1 + \cos(2\omega t + 2\theta_v)\right] + \frac{1}{2}V_m I_m \sin(\theta_v - \theta_i)\sin(2\omega t + 2\theta)$$

In this equation, the first term on the right-hand side is alwayg positive or zero,whereas the second term takes either positive or negative value,which illustrates the exchange of power in the circuit. The peak instantaneous power associated with the energy storage elements (i.e. the peak value of the second term on the right-hand side of the equation above) is called the reactive power, which is expressed by

$$Q = V_{rms}I_{rms}\sin\theta \tag{5.35}$$

Reactive power has the unit of Volt-ampere reactive (var).

Apparent power is defined as the product of the effective values of voltage and current

$$S = V_{rms}I_{rms} \tag{5.36}$$

and its unit is volt-ampere (VA).

From (5.33), (5.35) and (5.36), it can be seen that average power, reactive power and apparent power satisfy the following equation

$$P^2 + Q^2 = S^2 \tag{5.37}$$

功率，这说明功率三角形可以与阻抗三角形对应起来。阻抗 $Z = |Z| \angle \theta = R + jX$ 在复平面上的示意图如图 5.12 所示，其中 θ 同样反映的是电压与电流的相位差，即 $\theta = \theta_v - \theta_i$。根据图 5.12，显然有

$$\cos \theta = \frac{R}{|Z|} \tag{5.38}$$

$$\sin \theta = \frac{X}{|Z|} \tag{5.39}$$

由

$$I = \frac{V}{Z} = \frac{V_m \angle \theta_v}{|Z| \angle \theta} = \frac{V_m}{|Z|} \angle (\theta_v - \theta) \tag{5.40}$$

可知

$$I_m = \frac{V_m}{|Z|} \tag{5.41}$$

若用有效值表示，则有

$$I_{rms} = \frac{V_{rms}}{|Z|} \tag{5.42}$$

将式（5.38）和式（5.42）代入式（5.33），有

$$P = V_{rms} I_{rms} \frac{R}{|Z|} = I_{rms}^2 |Z| \frac{R}{|Z|} = I_{rms}^2 R \tag{5.43}$$

$$P = V_{rms} I_{rms} \frac{R}{|Z|} = V_{rms}^2 \frac{R}{R^2 + X^2} \tag{5.44}$$

将式（5.39）和式（5.42）代入式（5.35），可得

$$Q = V_{rms} I_{rms} \frac{X}{|Z|} = I_{rms}^2 |Z| \frac{X}{|Z|} = I_{rms}^2 X \tag{5.45}$$

$$Q = V_{rms} I_{rms} \frac{X}{|Z|} = V_{rms}^2 \frac{X}{R^2 + X^2} \tag{5.46}$$

【例 5.3】　求如图 5.13 所示电路中，电压源提供的平均功率。

解　首先计算电路总负载

$$Z = Z_{R1} + \frac{Z_{R2} Z_C}{Z_{R2} + Z_C} = 0.8 + \frac{4 \times (-j2)}{4 - j2} = 1.6\sqrt{2} \angle -45° \ \Omega$$

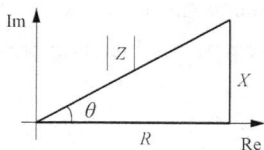

图 5.12　负载阻抗在复平面上的示意图

Fig. 5.12　Load impedance in the complex plane

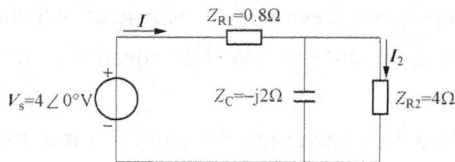

图 5.13　例 5.3 的电路图

Fig. 5.13　Circuit diagram for Example 5.3

进而求出电源电流

$$I = \frac{V_s}{Z} = \frac{4 \angle 0°}{1.6\sqrt{2} \angle -45°} = 1.25\sqrt{2} \angle 45° \text{A}$$

and they also form the power triangle as illustrated in Fig. 5.11.

The average power and reactive power of a circuit refer to the power absorbed by resistors and the power exchanged in the circuit by energy-storage elements (capacitors and inductors), respectively. Hence, the power triangle can be related to the impedance triangle. An impedance $Z = |Z|\angle\theta = R + jX$ is illustrated in the complex plane as shown in Fig. 5.12, where θ is the phase difference between the voltage and the current, i.e. $\theta = \theta_v - \theta_i$. According to this figure, we have

$$\cos\theta = \frac{R}{|Z|} \tag{5.38}$$

$$\sin\theta = \frac{X}{|Z|} \tag{5.39}$$

From

$$I = \frac{V}{Z} = \frac{V_m\angle\theta_v}{|Z|\angle\theta} = \frac{V_m}{|Z|}\angle(\theta_v - \theta) \tag{5.40}$$

we have

$$I_m = \frac{V_m}{|Z|} \tag{5.41}$$

If expressed using rms values, we have

$$I_{rms} = \frac{V_{rms}}{|Z|} \tag{5.42}$$

Substituting (5.38) and (5.42) into (5.33), we can write

$$P = V_{rms}I_{rms}\frac{R}{|Z|} = I_{rms}^2|Z|\frac{R}{|Z|} = I_{rms}^2 R \tag{5.43}$$

$$P = V_{rms}I_{rms}\frac{R}{|Z|} = V_{rms}^2\frac{R}{R^2 + X^2} \tag{5.44}$$

Similarly, substituting (5.39) and (5.42) into (5.35), we obtain

$$Q = V_{rms}I_{rms}\frac{X}{|Z|} = I_{rms}^2|Z|\frac{X}{|Z|} = I_{rms}^2 X \tag{5.45}$$

$$Q = V_{rms}I_{rms}\frac{X}{|Z|} = V_{rms}^2\frac{X}{R^2 + X^2} \tag{5.46}$$

Example 5.3: For the circuit shown in Fig. 5.13, find the average power provided by the voltage source.

Solution:

First we calculate the total load of the circuit

$$Z = Z_{R1} + \frac{Z_{R2}Z_C}{Z_{R2} + Z_C} = 0.8 + \frac{4\times(-j2)}{4 - j2} = 1.6\sqrt{2}\angle - 45°\,\Omega$$

根据式（5.31）电源提供的平均功率为

$$P = \frac{1}{2}V_{\mathrm{m}}I_{\mathrm{m}}\cos(\theta_v - \theta_i) = \frac{1}{2} \times 4 \times 1.25\sqrt{2}\cos(0 - 45°) = 2.5\ \mathrm{W}$$

从另一角度考虑这个问题，电源提供的平均功率即为各电阻消耗的功率之和。先计算流过电阻 R_2 的电流，根据分流原理

$$\boldsymbol{I}_2 = \frac{Z_{\mathrm{C}}}{Z_{\mathrm{C}} + Z_{\mathrm{R2}}}\boldsymbol{I} = \frac{-\mathrm{j}2}{-\mathrm{j}2 + 4} \times 1.25\sqrt{2}\angle 45° = \frac{2.5}{\sqrt{10}}\angle 45°\ \mathrm{A}$$

根据式（5.43），电阻 R_1 和 R_2 消耗的功率分别为

$$P_1 = I_{\mathrm{rms}}^2 R_1 = 1.25^2 \times 0.8 = 1.25\ \mathrm{W}$$

$$P_2 = I_{\mathrm{2rms}}^2 R_2 = \left(\frac{2.5}{\sqrt{20}}\right)^2 \times 4 = 1.25\ \mathrm{W}$$

则 $P = P_1 + P_2 = 2.5\ \mathrm{W}$，与第一种解法得到的结果一致。

✎ **讨论：** 功率三角形的实际应用——无功功率补偿

虽然无功功率并没有被消耗掉，但是它仍然在电路中进行交换。若无功功率过高，则电路中会有大电流流过，在电线上的能量损失会相应增加，而且电线所能承受的功率额定值需要远高于平均功率，这也造成了浪费。提高功率因数的方法为无功功率补偿。日常生活中的用电器多为感性的，为了提高功率因数，通常将一组电容与感性负载并联，让无功功率在电容与负载之间交换，而不要在整个电网中交换。

【**例 5.4**】　有一 50 kW 感性负载在额定频率 50 Hz、额定电压（有效值）10 kV 的电路中运行，功率因数为 0.6。为将功率因数提高至 0.9，需要并联多大的电容？

解　该感性负载的功率三角形如图 5.14（a）所示，其中 $P_1 = 50\ \mathrm{kW}$ 为已知，θ_1 可通过功率因数求出

$$\theta_1 = \arccos(0.6) = 53.13°$$

则无功功率为

$$Q_1 = P_1 \tan\theta_1 = 50\tan(53.13°) = 66.67\ \mathrm{kvar}$$

在并联电容后，功率因数提高至 0.9，则有

$$\theta_2 = \arccos(0.9) = \pm 25.84°$$

当 $\theta_2 = 25.84°$ 时，总负载仍为感性，所需并联的电容不需要太大。而若取 $\theta_2 = -25.84°$，则需要并联一个很大的电容将总负载的特性调整为容性，而功率因数仍为 0.9。因此选 $\theta_2 = 25.84°$ 即可。并联电容后的功率三角形如图 5.14（b）所示。

由于在感性负载两端并联了一个电容，并不影响总负载的平均功率，因此 $P_1 = P_2$。此时总负载的无功功率为

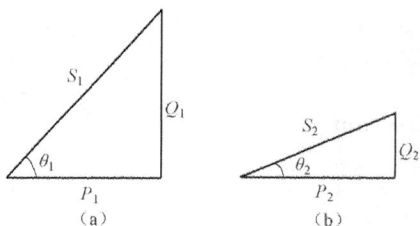

图 5.14　功率三角形

（a）感性负载的功率三角形；

（b）并联了补偿电容后的负载的功率三角形

Fig. 5.14　Power triangles

(a) The power triangle of the inductive load;

(b) The power triangle of the load after a

capacitor is connected in parallel to the inductor

$$Q_2 = P_2 \tan\theta_2 = 50\tan(25.84°) = 24.22\ \mathrm{kvar}$$

Then we solve for the source current

$$I = \frac{V_s}{Z} = \frac{4\angle 0°}{1.6\sqrt{2}\angle -45°} = 1.25\sqrt{2}\angle 45° \text{A}$$

According to (5.31), the average power provided by the source is

$$P = \frac{1}{2}V_m I_m \cos(\theta_v - \theta_i) = \frac{1}{2} \times 4 \times 1.25\sqrt{2}\cos(0 - 45°) = 2.5 \text{ W}$$

We may also consider this problem from another aspect. The average power provided by the source equals to the sum of the average power absorbed by the resistors. Let's first calculate the current flowing through R_2. Based on the current division principle, we have

$$I_2 = \frac{Z_C}{Z_C + Z_{R2}} I = \frac{-j2}{-j2 + 4} \times 1.25\sqrt{2}\angle 45° = \frac{2.5}{\sqrt{10}}\angle 45° \text{A}$$

According to (5.43), the average power absorbed by R_1 and R_2, respectively, is

$$P_1 = I_{1\text{rms}}^2 R_1 = 1.25^2 \times 0.8 = 1.25 \text{ W}$$

$$P_2 = I_{2\text{rms}}^2 R_2 = \left(\frac{2.5}{\sqrt{20}}\right)^2 \times 4 = 1.25 \text{ W}$$

Thus, we have $P = P_1 + P_2 = 2.5 \text{ W}$, which is consistent with the result we obtained previously.

Discussion: Application of power triangles—power factor correction

Reactive power is not absorbed by loads, but it does exchange within the circuit. A high reactive power also results in large current flowing in the circuit, which will increase the loss along the transmission line. Meanwhile, the lines must have higher ratings than what is necessary for average power, which is also a waste. A common practice to enhance power factor is reactive power compensation. Most appliances we use in our everyday life are inductive. Thus, to enhance power factor, we connect a group of capacitors in parallel with the inductive load, so that the reactive power is exchanged between the capacitors and the load, rather than in the whole electric network.

Example 5.4: A 50 kW inductive load operates in a circuit whose rated frequency is 50 Hz and rated voltage (effective value) is 10 kV. Its power factor is 0.6. Compute the capacitance that must be placed in parallel with the load in order to increase the power factor to 0.9.

Solution:

The power triangle of the inductive load is illustrated in Fig. 5.14(a), where $P_1 = 50 \text{ kW}$ is given and θ_1 can be calculated from the power factor

$$\theta_1 = \arccos(0.6) = 53.13°$$

Thus its reactive power is

$$Q_1 = P_1 \tan \theta_1 = 50 \times \tan(53.13°) = 66.67 \text{ kvar}$$

After a capacitor is connected in parallel, the power factor is raised to 0.9. Therefore, we have

$$\theta_2 = \arccos(0.9) = \pm 25.84°$$

When $\theta_2 = 25.84°$, the total load is still inductive, hence we only need to connect a small capacitor. Whereas if $\theta_2 = -25.84°$, a rather large capacitor is required, yet it does not help with the power factor. Therefore, we choose $\theta_2 = 25.84°$. After adding the capacitor, the power triangle of the total load is shown in Fig. 5.14(b).

因此，电容的无功功率为

$$Q_C = Q_2 - Q_1 = 24.22 - 66.67 = -42.45 \text{ kvar}$$

由于线路额定电压有效值为 10 kV，即电容两端电压为 10 kV，则根据式（5.46）可知，电容的电抗为

$$X_C = \frac{V_{\text{rms}}^2}{Q_C} = \frac{(10 \times 10^3)^2}{-42.45 \times 10^3} = -2356 \, \Omega$$

进而求得电容为

$$C = -\frac{1}{\omega X_C} = -\frac{1}{2\pi \times 50 \times (-2356)} = 1.35 \, \mu\text{F}$$

As adding a capacitor does not affect the average power, we have $P_1 = P_2$. And the reactive power of the total load can be calculated from

$$Q_2 = P_2 \tan \theta_2 = 50 \tan(25.84°) = 24.22 \text{ kvar}$$

Thus the reactive power of the capacitor must be

$$Q_C = Q_2 - Q_1 = 24.22 - 66.67 = -42.45 \text{kvar}$$

As the power line has a rated rms voltage of 10 kV, the voltage across the capacitor is 10 kV. According to (5.46), the reactance of the capacitor is

$$X_C = \frac{V_{rms}^2}{Q_C} = \frac{(10 \times 10^3)^2}{-42.45 \times 10^3} = -2356 \ \Omega$$

and the required capacitance is

$$C = -\frac{1}{\omega X_C} = -\frac{1}{2\pi \times 50 \times (-2356)} = 1.35 \ \mu\text{F}$$

▶ Problems

P5.1   For the circuit shown in Fig. P5.1, where $v_s = 100\cos(200t)$ V, $R = 100 \ \Omega$, $L = 50 \text{ mH}$, $C = 0.2 \ \mu\text{F}$, find the phasors V_s, V_R, V_C, V_L and I.

P5.2   Find the node voltages V_1 and V_2 in phasors in the circuit shown in Fig. P5.2.

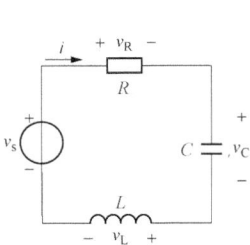

Fig. P5.1

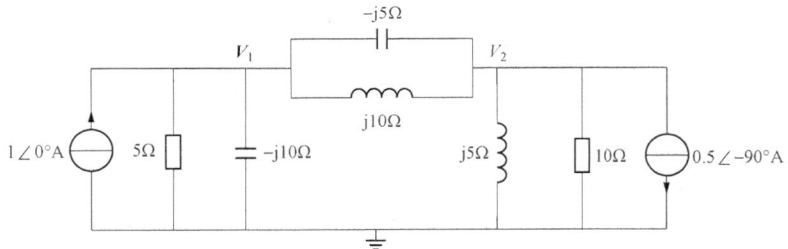

Fig. P5.2

P5.3   Solve for current I in the circuit shown in Fig. P5.3, where $V_{s1} = 100\angle 0°$ V, $I_{s2} = 2\angle 90°$ A, $R = 50 \ \Omega$, $X_L = 50 \ \Omega$, $X_C = 100 \ \Omega$.

P5.4   Consider the circuit shown in Fig. P5.4, where $I_R = I_L = 10$ A, $V = 100$ V, V and I have the same phase angle. Suppose that the phase of V_R is $0°$. Calculate I, R, X_L and X_C.

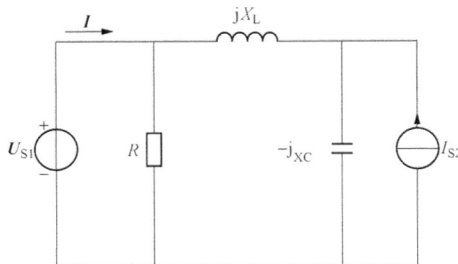

Fig. P5.3

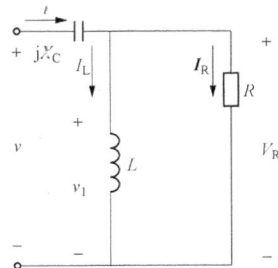

Fig. P5.4

117

P5.5　Find the Thévenin and Norton equivalent circuits for the circuit shown in Fig. P5.5.

P5.6　For the circuit shown in Fig. P5.6, where $v = 400\sqrt{2}\sin(1000t)$ V, the branch of R_1 has power $P_1 = 2.5$ kW, i_1 and v have the same phase, $R_2 = 40\,\Omega$, and $C_2 = 25\,\mu\text{F}$. Find out R_1, v_1, i_1, i_2 and i.

P5.7　In the power system shown in Fig. P5.7, the voltage source generates a voltage of 1 kV in rms value with the frequency of 50 Hz. Two loads are connected in parallel to the source. Load A consumes 10 kW with a 90% lagging power factor. Load B has an apparent power of 15 kVA with an 80% lagging power factor. Find the power, reactive power, and apparent power delivered by the source. What is the power factor seen by the source?

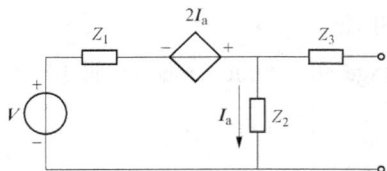

Fig. P5.5

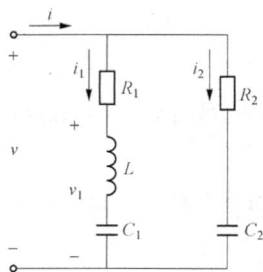

Fig. P5.6

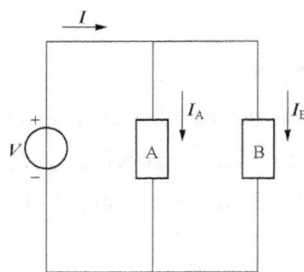

Fig. P5.7

第6章
Chapter 6

三 相 电 路
Three–phase Circuits

6.1 三相电路的基本概念

之前讨论的都是单相电路，电路中有一个单相电源通过导线向负载传输电能。但在实际的电力系统中，人们普遍采用的是三相电路。在这种电路中，三相发电机产生三个正弦电压，它们的幅值和频率都相同，相位相差 120°，如图 6.1 所示。这三个正弦电压依次称为 A 相、B 相和 C 相，它们的表达式分别为

$$v_{an}(t) = V\cos(\omega t) \tag{6.1}$$

$$v_{bn}(t) = V\cos(\omega t - 120°) \tag{6.2}$$

$$v_{cn}(t) = V\cos(\omega t - 240°) = V\cos(\omega t + 120°) \tag{6.3}$$

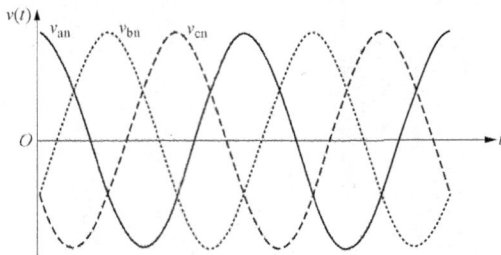

图 6.1 三相电压波形示意图，各相电压相位相差 120°

Fig. 6.1 An illustration of three-phase voltages which are 120° apart from each other

它们对应的相量形式分别为

$$V_{an} = V_Y \angle 0° \tag{6.4}$$

$$V_{bn} = V_Y \angle -120° \tag{6.5}$$

$$V_{cn} = V_Y \angle 120° \tag{6.6}$$

上述符合 A、B、C 顺序的三相电称为正序三相电，即 A 相电压领先 B 相电压 120°，B 相电压超前 C 相电压 120°。若三相电按 A、C、B 顺序排列，即 A 相领先 C 相 120°，C 相超前 B 相 120°，则称为逆序三相电。

6.2 三相电源与负载的星形和三角形接法

6.2.1 Y-Y 连接

在图 6.2 中，三相电源和三个负载均按星形方式连接，称为 Y-Y 连接。线路 aA、bB、cC 称为端线，线路 nN 称为中性线。三个电压源满足式（6.1）～式（6.3）且三个负载阻抗相同，这样的电路称为对称三相电路。而在实际中，三相电源是对称的，而负载阻抗则不一定相同，这样的电路称为不对称三相电路。在本书中我们只考虑对称三相电路。

各端线与中线间的电压称为相电压，在 Y-Y 连接的情况下相电压也是电源电压，其表达式为式（6.1）～式（6.3）或式（6.4）～式（6.6）。两端线之间的电压称为线电压，分别记为 V_{ab}、V_{bc} 和 V_{ca}，根据 KVL 有

$$V_{ab} = V_{an} - V_{bn} = V_Y \angle 0° - V_Y \angle -120° = \sqrt{3}V_Y \angle 30° = V_{an} \cdot \sqrt{3}\angle 30° \tag{6.7}$$

$$V_{bc} = V_{bn} - V_{cn} = \sqrt{3}V_Y \angle -90° = V_{bn} \cdot \sqrt{3}\angle 30° \tag{6.8}$$

6.1   Basic Concepts

In the previous chapters, we dealt with single-phase circuits consisting of a generator for providing electricity to the loads via a pair of wires. In practical power systems, what people normally use are three-phase circuits. In such a circuit, a generator generates three source voltages, which have the same amplitude and frequency but are out of phase with each other by 120°. As shown in Fig. 6.1, the expressions for the three voltages are

$$v_{an}(t) = V\cos(\omega t) \tag{6.1}$$

$$v_{bn}(t) = V\cos(\omega t - 120°) \tag{6.2}$$

$$v_{cn}(t) = V\cos(\omega t - 240°) = V\cos(\omega t + 120°) \tag{6.3}$$

To express them in terms of phasors, we have

$$V_{an} = V_Y\angle 0° \tag{6.4}$$

$$V_{bn} = V_Y\angle -120° \tag{6.5}$$

$$V_{cn} = V_Y\angle 120° \tag{6.6}$$

Such an arrangement of the voltages is called a positive phase sequence, as the voltage of phase A leads that of phase B by 120° and the voltage of phase B leads that of phase C by 120°. If the voltages come in the order of a-b-c, i.e. the voltage of phase A leads that of phase C by 120°, and the voltage of phase C leads that of phase B by 120°, we say they have a negative phase sequence.

6.2   The Wye–Wye Connection and Delta–Delta Connection

6.2.1   Wye-wye connection

Consider the three-phase circuit shown in Fig. 6.2, where both the sources and loads are connected forming the shape of letter 'Y'. Such a configuration is called wye-wye connection. The wires of a-A, b-B and c-C are called lines, and the wire of n-N is called the neutral. The three voltage sources satisfy (6.1)~(6.3), and the three loads are of the same impedance. Such a circuit is called a balanced three-phase circuit. In most practical cases, the three-phase sources are usually balanced, whereas the loads may not be. Thus, we have unbalanced three-phase circuits. In this book, we consider only the balanced ones.

The voltage between each line and the neutral is called line-to-neutral voltage or phase voltage, which is also the source voltage in a wye-wye connected circuit; so we can use (6.1)~(6.3) or (6.4)~(6.6) to express them. The voltage between each pair of lines, i.e. the voltage between node a and node b, node b and node c, or node c and node a, is called line-to-line voltage, or simply line voltage. Denote the voltages by V_{ab}, V_{bc} and V_{ca}, respectively. According to KVL, we have

$$V_{ab} = V_{an} - V_{bn} = V_Y\angle 0° - V_Y\angle -120° = \sqrt{3}V_Y\angle 30° = V_{an} \cdot \sqrt{3}\angle 30° \tag{6.7}$$

$$V_{bc} = V_{bn} - V_{cn} = V_{bn} \cdot \sqrt{3}\angle 30° \tag{6.8}$$

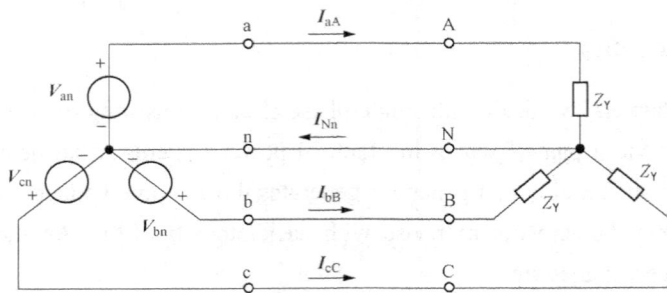

图 6.2　三相电源与负载的星形接法

Fig. 6.2　Wye-wye connection

$$V_{ca} = V_{cn} - V_{an} = \sqrt{3}V_Y \angle 150° = V_{cn} \cdot \sqrt{3} \angle 30° \qquad (6.9)$$

记线电压幅值为 V_L，则有 $V_L = \sqrt{3}V_Y$。

各电压源流出的电流称为相电流，各端线中的电流称为线电流。在 Y-Y 连接的情况下，相电流与线电流相同，均可通过下式计算

$$I_{aA} = \frac{V_{an}}{Z_Y} = \frac{V_Y \angle 0°}{|Z_Y| \angle \theta} = I_L \angle -\theta \qquad (6.10)$$

$$I_{bB} = \frac{V_{bn}}{Z_Y} = I_L \angle (-120° - \theta) \qquad (6.11)$$

$$I_{cC} = \frac{V_{cn}}{Z_Y} = I_L \angle (120° - \theta) \qquad (6.12)$$

式中，$I_L = V_Y / Z_Y$ 为线电流的幅值。在中性点处应用 KCL 可以得到

$$I_{Nn} = I_{aA} + I_{bB} + I_{cC} = 0 \qquad (6.13)$$

由此可见，中性线中没有电流流过，因此可以将其去掉。带有中性线的 Y-Y 连接称为三相四线制方式，其他连接方式均为三相三线制。如果将上述三个电压源和三个负载分别连接成三个独立的单相电路，则需要六条线路。而采用三相电路，则只需要三条或四条线路。在实际电力系统中，输电距离通常为几十、几百、甚至上千公里，采用三相电路可以极大地节约成本。

6.2.2　三相电路的功率

首先计算如图 6.2 所示的三相电路的瞬时功率

$$\begin{aligned}
p(t) &= v_{an}(t)i_{aA}(t) + v_{bn}(t)i_{bB}(t) + v_{cn}(t)i_{cC}(t) \\
&= V_Y\cos(\omega t)I_L\cos(\omega t - \theta) + V_Y\cos(\omega t - 120°)I_L\cos(\omega t - 120° - \theta) \\
&\quad + V_Y\cos(\omega t + 120°)I_L\cos(\omega t + 120° - \theta) \qquad (6.14) \\
&= 3\frac{V_Y I_L}{2}\cos\theta
\end{aligned}$$

这是一个常数，不随时间而变化，这是三相电路优于单相电路的另一个优势。显然电路的平均功率等于其瞬时功率。将相电压和线电流的有效值代入式（6.14），可得

$$P = p(t) = 3V_{Yrms}I_{Lrms}\cos\theta \qquad (6.15)$$

三相对称电路的无功功率计算方法与第 5 章所述相同，为

$$Q = 3V_{Yrms}I_{Lrms}\sin\theta \qquad (6.16)$$

$$V_{ca} = V_{cn} - V_{an} = V_{cn} \cdot \sqrt{3} \angle 30° \tag{6.9}$$

If we use V_L to indicate the amplitude of line voltages, we have $V_L = \sqrt{3}V_Y$.

Currents flowing out of the sources are called phase currents, while currents flowing through the lines are called line currents. In a wye-wye connected circuit, phase currents are the same as the corresponding line currents, which can be calculated from the following equations

$$\boldsymbol{I}_{aA} = \frac{\boldsymbol{V}_{an}}{Z_Y} = \frac{V_Y \angle 0°}{|Z_Y| \angle \theta} = I_L \angle -\theta \tag{6.10}$$

$$\boldsymbol{I}_{bB} = \frac{\boldsymbol{V}_{bn}}{Z_Y} = I_L \angle(-120° - \theta) \tag{6.11}$$

$$\boldsymbol{I}_{cC} = \frac{\boldsymbol{V}_{cn}}{Z_Y} = I_L \angle(120° - \theta) \tag{6.12}$$

where $I_L = V_Y / Z_Y$ denotes the amplitude of line currents. Applying KCL at the neutral point n or N, we have

$$\boldsymbol{I}_{Nn} = \boldsymbol{I}_{aA} + \boldsymbol{I}_{bB} + \boldsymbol{I}_{cC} = 0 \tag{6.13}$$

The above equation implies that no current flows through the neutral, which can thus be removed. A wye-wye connection with the neutral is called a three-phase four-line system, while other connections are three-phase three-line systems. If we connect the voltage sources and the loads in three independent single-phase circuits, we will need six wires; whereas if three-phase circuits are applied, only three or four wires are required. In real power systems, electricity is transmitted for tens, hundreds, or thousands of kilometers. In this case, using three-phase circuits is a much more economical choice.

6.2.2　Power in three-phase circuits

We start from calculating the instantaneous power of the three-phase circuit shown in Fig. 6.2

$$\begin{aligned} p(t) &= v_{an}(t)i_{aA}(t) + v_{bn}(t)i_{bB}(t) + v_{cn}(t)i_{cC}(t) \\ &= V_Y \cos(\omega t)I_L \cos(\omega t - \theta) + V_Y \cos(\omega t - 120°)I_L \cos(\omega t - 120° - \theta) \\ &\quad + V_Y \cos(\omega t + 120°)I_L \cos(\omega t + 120° - \theta) \\ &= 3\frac{V_Y I_L}{2} \cos\theta \end{aligned} \tag{6.14}$$

Apparently, the instantaneous power is a constant that does not change with time. This is another advantage of three-phase circuits over single-phase ones. In this case, the expression of power is exactly the same as that of the instantaneous power. If we use the rms values in (6.14), it becomes

$$P = p(t) = 3V_{Yrms}I_{Lrms}\cos\theta \tag{6.15}$$

For the calculation of reactive power, we apply the same strategy as used in Chapter 5

$$Q = 3V_{Yrms}I_{Lrms}\sin\theta \tag{6.16}$$

6.2.3 △-△连接

三相电路的另外一种常用接法为△-△连接，如图 6.3 所示，三个电压源和三个负载（阻抗均为 $Z_\triangle = |Z_\triangle| \angle \theta$）分别接成三角形。这里电源电压 V_{ab}、V_{bc} 和 V_{ca} 分别等于各线电压。将图 6.3 所示电路与图 6.2 所示电路对应起来，有

$$V_{ab} = V_L \angle 30° \tag{6.17}$$

$$V_{bc} = V_L \angle -90° \tag{6.18}$$

$$V_{ca} = V_L \angle 150° \tag{6.19}$$

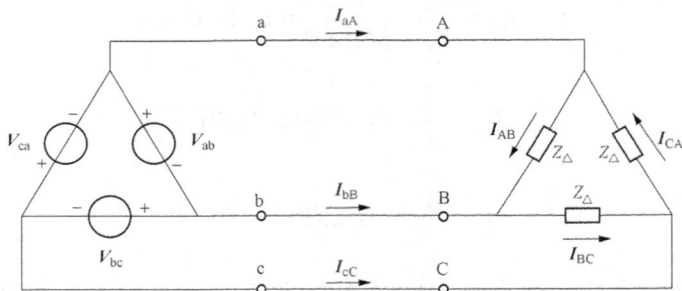

图 6.3　三相电源与负载的三角形接法

Fig. 6.3　△-△ connection

各负载电流分别为

$$I_{AB} = \frac{V_{ab}}{Z_\triangle} = \frac{V_L \angle 30°}{|Z_\triangle| \angle \theta} = I_\triangle \angle (30° - \theta) \tag{6.20}$$

$$I_{BC} = \frac{V_{bc}}{Z_\triangle} = \frac{V_L \angle -90°}{|Z_\triangle| \angle \theta} = I_\triangle \angle (-90° - \theta) \tag{6.21}$$

$$I_{CA} = \frac{V_{ca}}{Z_\triangle} = \frac{V_L \angle 150°}{|Z_\triangle| \angle \theta} = I_\triangle \angle (150° - \theta) \tag{6.22}$$

式中，$I_\triangle = V_L / |Z_\triangle|$ 为负载电流幅值。分别在节点 A、B、C 处应用 KCL，可得线电流与负载电流之间的关系为

$$I_{aA} = I_{AB} - I_{CA} = I_{AB} \cdot \sqrt{3} \angle -30° \tag{6.23}$$

$$I_{bB} = I_{BC} - I_{AB} = I_{BC} \cdot \sqrt{3} \angle -30° \tag{6.24}$$

$$I_{cC} = I_{CA} - I_{BC} = I_{CA} \cdot \sqrt{3} \angle -30° \tag{6.25}$$

线电流幅值与负载电流幅值之间关系为 $I_L = \sqrt{3} I_\triangle$。

6.3 三相电路分析

三相电路实际上是一种特殊类型的正弦电路，因此可以将在第 5 章中学到的正弦稳态电路分析方法应用至三相电路分析中。另外，三相电路的对称性也为求解提供了方便。

【例 6.1】　在如图 6.4 所示的对称三相三线制电路中，已知电源电压有效值为 220 V，频率为 50 Hz，各相负载为 0.2 H 电感与 50 Ω 电阻的串联。假设这是一个正序三相且 A 相电压相位为 0°。求各线电压、线电流、负载消耗的平均功率和无功功率。

6.2.3   Delta-delta connection

Delta-delta connection is also commonly used in three-phase circuits. As shown in Fig. 6.3, the three voltage sources and the three loads are connected into a triangular configuration, respectively.

The loads have the same impedance of $Z_\triangle = |Z_\triangle| \angle \theta$. Under such configuration, the source voltages V_{ab}, V_{bc} and V_{ca} are actually line voltages. Suppose Fig. 6.2 and Fig. 6.3 share the same line voltages, we have

$$V_{ab} = V_L \angle 30° \tag{6.17}$$

$$V_{bc} = V_L \angle -90° \tag{6.18}$$

$$V_{ca} = V_L \angle 150° \tag{6.19}$$

The load currents are calculated from

$$I_{AB} = \frac{V_{ab}}{Z_\triangle} = \frac{V_L \angle 30°}{|Z_\triangle| \angle \theta} = I_\triangle \angle(30° - \theta) \tag{6.20}$$

$$I_{BC} = \frac{V_{bc}}{Z_\triangle} = \frac{V_L \angle -90°}{|Z_\triangle| \angle \theta} = I_\triangle \angle(-90° - \theta) \tag{6.21}$$

$$I_{CA} = \frac{V_{ca}}{Z_\triangle} = \frac{V_L \angle 150°}{|Z_\triangle| \angle \theta} = I_\triangle \angle(150° - \theta) \tag{6.22}$$

where $I_\triangle = V_L / |Z_\triangle|$ represents the amplitude of load currents. Applying KCL at nodes A, B and C, respectively, we obtain the following relationships between line currents and load currents

$$I_{aA} = I_{AB} - I_{CA} = I_{AB} \cdot \sqrt{3} \angle -30° \tag{6.23}$$

$$I_{bB} = I_{BC} - I_{AB} = I_{BC} \cdot \sqrt{3} \angle -30° \tag{6.24}$$

$$I_{cC} = I_{CA} - I_{BC} = I_{CA} \cdot \sqrt{3} \angle -30° \tag{6.25}$$

For the amplitude of line currents and that of load currents, we have $I_L = \sqrt{3} I_\triangle$.

6.3   Analysis of Balanced Three–phase Circuits

Three-phase circuits are actually a special type of sinusoidal circuits. Therefore, we can apply the steady-state sinusoidal analysis techniques to three-phase circuit analysis. Besides, it is convenient to solve the circuits as they are balanced.

Example 6.1: Consider the balanced three-phase circuit shown in Fig. 6.4, which is a three-phase three-line system. The sources provide sinusoidal voltages of 220 V amplitude (rms value) and 50 Hz frequency. For each phase the load is a 0.2 H inductor connected in series with a 50 Ω resistor. Suppose the source voltages have a positive phase sequence and the phase angle corresponding to the voltage of phase A is 0°. Solve for the line voltages, line currents, average power and reactive power delivered to the loads.

Solution:

According to the given information, we can write the phase voltages

$$V_{an} = 220\sqrt{2} \angle 0° \text{V}$$

$$V_{bn} = 220\sqrt{2} \angle -120° \text{V}$$

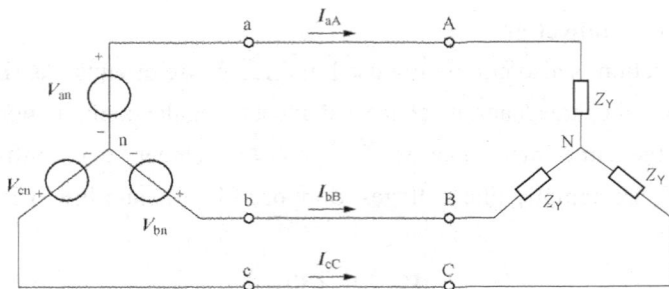

图 6.4 例 6.1 的电路图

Fig. 6.4 Circuit diagram for Example 6.1

解 根据已知条件，可得到各相电压为

$$V_{an} = 220\sqrt{2}\angle 0° \text{ V}$$

$$V_{bn} = 220\sqrt{2}\angle -120° \text{ V}$$

$$V_{cn} = 220\sqrt{2}\angle 120° \text{ V}$$

由于这是一个对称三相电路，可以假设节点 n 和节点 N 是相连的。根据式（6.7）～式（6.9），可求得各线电压为

$$V_{ab} = V_{an} \cdot \sqrt{3}\angle 30° = 381\sqrt{2}\angle 30° \text{ V}$$

$$V_{bc} = V_{bn} \cdot \sqrt{3}\angle 30° = 381\sqrt{2}\angle -90° \text{ V}$$

$$V_{ca} = V_{cn} \cdot \sqrt{3}\angle 30° = 381\sqrt{2}\angle 150° \text{ V}$$

由已知条件也可知

$$Z_Y = R + j\omega L = 50 + j2\pi \times 50 \times 0.2 = 50 + j62.8 = 80.27\angle 51.47° \text{ }\Omega$$

根据欧姆定律和式（6.10）～式（6.12），各线电流为

$$I_{aA} = \frac{V_{an}}{Z_Y} = \frac{220\sqrt{2}\angle 0°}{80.27\angle 51.47°} = 3.88\angle -51.47° \text{A}$$

$$I_{bB} = I_L\angle -120° -51.47° = 3.88\angle -171.47° \text{ A}$$

$$I_{cC} = I_L\angle 120° -51.47° = 3.88\angle 68.53° \text{ A}$$

负载消耗的平均功率和无功功率可分别由式（6.15）和式（6.16）求出：

$$P = 3V_{Yrms}I_{Lrms}\cos\theta = 3 \times 220 \times \frac{3.88}{\sqrt{2}}\cos(51.47°) = 1128.14 \text{ W}$$

$$Q = 3V_{Yrms}I_{Lrms}\sin\theta = 3 \times 220 \times \frac{3.88}{\sqrt{2}}\sin(51.47°) = 1001.63 \text{ var}$$

讨论： 三相电路的实际应用

三相电路的一个最普遍应用是电力系统。发电厂生产出三相交流电，经过输电线将电能送至用户侧供负载使用。由于电能通常为长距离传输，因此传输线上的损耗不能忽略不计，即在进行电路分析时需要考虑传输线的阻抗。

【例 6.2】 考虑如图 6.5（a）所示电路。这是一个△-△连接的对称三相电路，线电压 $V_{ab} = 14.14\sqrt{3}\angle 30° \text{ kV}$，导线阻抗 $Z_{line} = 0.3 + j0.4 \text{ }\Omega$，负载阻抗 $Z_\triangle = 30 + j6 \text{ }\Omega$。求线电流、负载处的线电压、负载消耗的平均功率和传输线上损失的平均功率。

$$V_{cn} = 220\sqrt{2}\angle 120° \text{ V}$$

As it is a balanced three-phase circuit, we can assume that node n and node N are connected. Hence, according to (6.7)~(6.9), the line voltages can be obtained from

$$V_{ab} = V_{an} \cdot \sqrt{3}\angle 30° = 381\sqrt{2}\angle 30° \text{ V}$$

$$V_{bc} = V_{bn} \cdot \sqrt{3}\angle 30° = 381\sqrt{2}\angle -90° \text{ V}$$

$$V_{ca} = V_{cn} \cdot \sqrt{3}\angle 30° = 381\sqrt{2}\angle 150° \text{ V}$$

The impedance of each phase of the loads is computed as

$$Z_Y = R + j\omega L = 50 + j2\pi \times 50 \times 0.2 = 50 + j62.8 = 80.27\angle 51.47° \text{ }\Omega$$

Thus, according to Ohm's Law and (6.10)~(6.12), the line currents are calculated as follows

$$I_{aA} = \frac{V_{an}}{Z_Y} = \frac{220\sqrt{2}\angle 0°}{80.27\angle 51.47°} = 3.88\angle -51.47° \text{ A}$$

$$I_{bB} = I_L\angle(-120° - 51.47°) = 3.88\angle -171.47° \text{ A}$$

$$I_{cC} = I_L\angle(120° - 51.47°) = 3.88\angle 68.53° \text{ A}$$

The power and reactive power delivered to the loads are given by (6.15) and (6.16)

$$P = 3V_{Yrms}I_{Lrms}\cos\theta = 3 \times 220 \times \frac{3.88}{\sqrt{2}}\cos(51.47°) = 1128.14 \text{ W}$$

$$Q = 3V_{Yrms}I_{Lrms}\sin\theta = 3 \times 220 \times \frac{3.88}{\sqrt{2}}\sin(51.47°) = 1001.63 \text{ var}$$

✓ **Discussion:**   Application of three-phase circuits in practice

Three-phase circuits are widely used in our power systems. Three-phase electricity is generated in power plants, transmitted through transmission lines to users, and consumed by loads. As electricity is usually transmitted through a long distance, the impedance of transmission lines should not be neglected. Therefore, we need to consider the line impedance in our analysis.

Example 6.2: Consider the circuit shown in Fig. 6.5 (a), which shows a balanced three-phase circuit in delta-delta connection. The line voltage is $V_{ab} = 14.14\sqrt{3}\angle 30° \text{ kV}$, each line has an impedance of $Z_{line} = 0.3 + j0.4 \text{ }\Omega$ and the load impedance of each phase is $Z_\triangle = 30 + j6 \text{ }\Omega$. Find the line current, line-to-line voltage at the load, the power consumed by the loads, and the power dissipated in the wires.

Nikola Tesla (10 July 1856 - 7 January 1943) was a Serbian American inventor, electrical engineer, mechanical engineer, and futurist. He is best known for his contributions to the design of the modern alternating current (ac) electricity supply system. The unit of magnetic flux density was named the tesla in his honor.

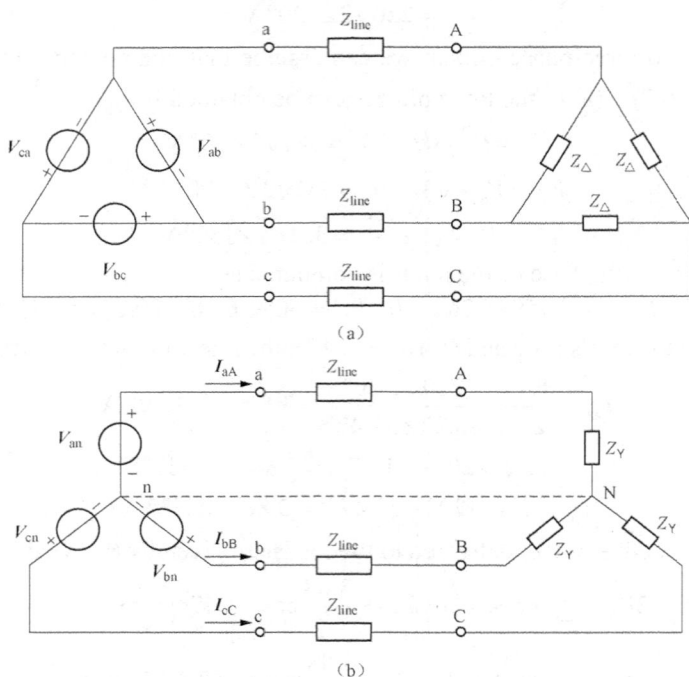

图 6.5　例 6.2 的电路图

（a）原始△-△连接的对称三相电路；（b）转换为Y-Y连接的等效电路

Fig. 6.5　Circuit diagram for Example 6.2

(a) Original balanced three-phase circuit in delta-delta connection;

(b) Equivalent circuit in wye-wye connection

解　线路存在阻抗，导致电源侧的线电压 V_{ab} 不再等于负载侧的线电压 V_{AB}。这种情况下继续基于△-△连接的电路进行求解将面临非常庞大的计算量。因此我们将△-△连接的电路等效转换成Y-Y连接的形式，如图6.5（b）所示。这样，线路阻抗不变，电源相电压可通过式（6.7）求得

$$V_{an} = \frac{V_{ab}}{\sqrt{3}\angle 30°} = 14.14\angle 0° \text{ kV}$$

根据第 2 章中电阻的等效变换公式（2.27）可知，△连接的负载在转化为 Y 连接后，其等效阻抗变为

$$Z_Y = Z_\triangle / 3 = 10 + j2 = 10.2\angle 11.31° \ \Omega$$

在 Y-Y 连接的对称三相电路中，可以假设中性线 nN 存在，则在 A 相回路中可根据欧姆定律列出如下方程

$$I_{aA} = \frac{V_{an}}{Z_L + Z_Y} = \frac{14.14\angle 0°}{0.3 + j0.4 + 10 + j2} = 1.34\angle -13.12° \text{ kA}$$

负载侧的相电压为

$$V_{AN} = I_{aA} Z_Y = 1.34\angle -13.12° \times 10.2\angle 11.31° = 13.67\angle -1.81° \text{ kV}$$

负载侧的线电压为

$$V_{AB} = V_{AN} \cdot \sqrt{3}\angle 30° = 23.68\angle 28.19° \text{ kV}$$

Solution:

As the impedance of the wires cannot be neglected, the line voltage at the source, V_{ab}, no longer equals the line voltage at the load, V_{AB}. Under such a delta-delta configuration, it takes a great deal of computation to solve the circuit. Therefore, we propose to transform the delta-delta connection into the equivalent wye-wye connection, as shown in Fig. 6.5(b). In this case, line impedance remains the same; phase voltage of the source can be calculated from

$$V_{an} = \frac{V_{ab}}{\sqrt{3}\angle 30°} = 14.14\angle 0° \text{ kV}$$

In Chapter 2 we learnt how to transform delta-connected resistances into its equivalent wye-connected resistances. In the same manner, we convert the delta-connected loads into wye connection, and the equivalent impedance is

$$Z_Y = Z_\triangle / 3 = 10 + j2 = 10.2\angle 11.31° \ \Omega$$

In a balanced three-phase circuit of wye-wye connection, we may assume the neutral points n and N are linked together. Hence, applying Ohm's Law in the loop of phase A, we get

$$I_{aA} = \frac{V_{an}}{Z_L + Z_Y} = \frac{14.14\angle 0°}{0.3 + j0.4 + 10 + j2} = 1.34\angle -13.12° \text{ kA}$$

The phase voltage at the load is

$$V_{AN} = I_{aA} Z_Y = 1.34\angle -13.12° \times 10.2\angle 11.31° = 13.67\angle -1.81° \text{ kV}$$

And the line voltage at the load is

$$V_{AB} = V_{AN} \bullet \sqrt{3}\angle 30° = 23.68\angle 28.19° \text{ kV}$$

Thomas Alva Edison (11 February, 1847 – 18 October, 1931) was an American inventor and businessman. He developed many devices that greatly influenced life around the world, including the phonograph, the motion picture camera, and a long-lasting, practical electric light bulb. Edison invented an entire electrical system of sockets, cables, meters to go with it, and a ground-breaking new way of distributing electricity. However, he was convinced by his dc low voltage system, hereby drew a battle line between dc and ac power systems, between Edison and Tesla.

由于电源提供的功率会被线路阻抗消耗一部分，因此在通过式（6.14）计算负载功率时，需要代入负载侧的相电压和线电流，即

$$P = 3\frac{V_Y I_L}{2}\cos\theta = 3\times\frac{|\boldsymbol{V}_{AN}||\boldsymbol{I}_{aA}|}{2}\cos(11.31°) = 26.94\ \text{kW}$$

A 相线路上损耗的功率为

$$P_{\text{Aline}} = I_{\text{aArms}}^2 R_{\text{line}} = \left(\frac{1.34}{\sqrt{2}}\right)^2 \times 0.3 = 0.28\ \text{kW}$$

其他两条线路上的损耗相同，因此总的线路损耗为

$$P_{\text{line}} = 3P_{\text{Aline}} = 0.85\ \text{kW}$$

It should be noted that the power supplied by the source is partly absorbed by the line. As a result, when we apply (6.14) to calculate the power delivered to the load, we need to use the phase voltage and line current at the load

$$P = 3\frac{V_Y I_L}{2}\cos\theta = 3\times\frac{|V_{AN}||I_{aA}|}{2}\cos(11.31°) = 26.94 \text{ kW}$$

The power lost in line a-A can be found by

$$P_{\text{Aline}} = I_{\text{aArms}}^2 R_{\text{line}} = \left(\frac{1.34}{\sqrt{2}}\right)^2 \times 0.3 = 0.28 \text{ kW}$$

The power dissipated in the other two lines is the same; thus, the total loss is

$$P_{\text{line}} = 3P_{\text{Aline}} = 0.85 \text{ kW}$$

▶ Problems

P6.1   A balanced positive-sequence wye-connected three-phase source has line-to-line voltages of 380 V rms. This source is connected to a balanced wye-connected load, as shown in Fig. P6.1. Each phase of the load is $Z = 3 - j4\,\Omega$. Find the instantaneous current of i_{aA}, i_{bB}, i_{cC}, i_{Nn}, the apparent power, average power and the reactive power delivered to the load. Assume that the phase of V_{ab} is zero.

P6.2   In the balanced three-phase system with a line-to-line voltage of 380 V in Fig. P6.2, each load has an impedance of $Z=24+j18\,\Omega$. Find the line current of I_{aA}, I_{bB}, I_{cC} and plot phasor diagram. Assume that the phasor of $V_{ab} = 380\angle 0°$.

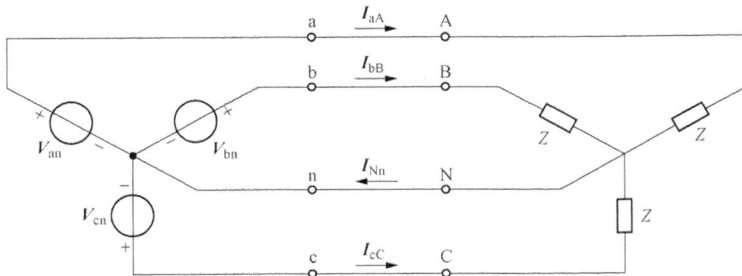

Fig. P6.1

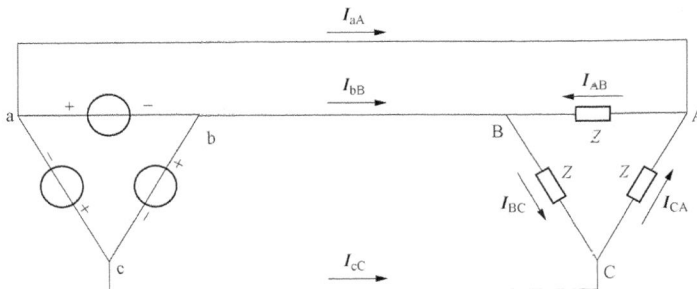

Fig. P6.2

二 极 管
Diodes

本章所介绍的二极管即晶体二极管，是一种半导体器件。二极管具有单向导电性，即电流从阳极流至阴极时可以很大，但从阴极流向阳极时却非常小。利用二极管这一特性可实现开关、整流和限幅等功能。同时二极管的其他特性还可用于解调、稳压和发光等，是一种应用广泛的电子器件。

本章首先从二极管的基本概念开始，介绍半导体材料的结构和 PN 结的特性，然后介绍数种用于电路分析的二极管模型，在此基础上，最后讨论了整流、限幅、箝位和斩波等几种二极管的实际应用电路。

7.1 二极管的基本概念

7.1.1 本征半导体和杂质半导体

半导体是导电性能介于导体和绝缘体之间的材料，常见的半导体材料是四价元素，如锗（Ge）和硅（Si）。由这些元素原子形成的结构纯净的单晶体就是本征半导体。此时晶体中每个原子外层的四个电子分别与相邻原子的一个外层电子成为共价键，形成稳定的结构，如图7.1（a）所示。在常温下，本征半导体中仅有少数外层电子能挣脱共价键束缚形成自由电子，因此本征半导体导电性能差。

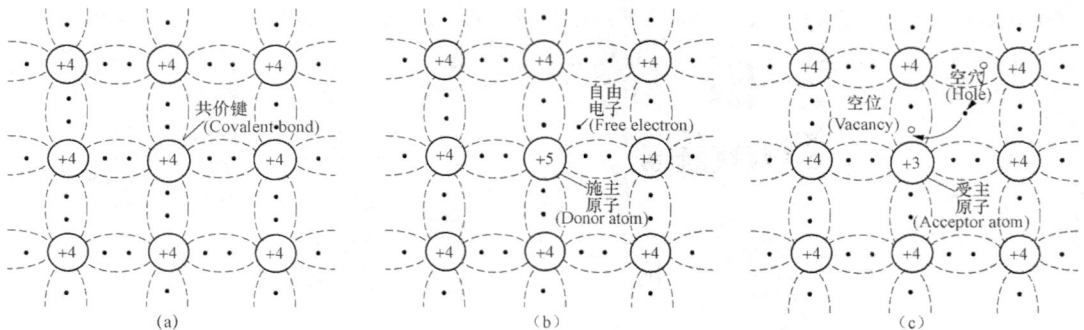

图 7.1　半导体材料结构

（a）本征半导体；（b）N 型半导体；（c）P 型半导体

Fig. 7.1　Structures of the semiconductor material

(a) Intrinsic semiconductor; (b) N-type semiconductor; (c) P-type semiconductor

本征半导体在掺入少量杂质元素后成为杂质半导体。如向硅本征半导体中掺入少量五价元素磷（P）后，磷原子外层的五个电子中，只有四个能与相邻硅原子的价电子形成共价键，剩下一个电子即使在常温下也容易成为自由电子，如图 7.1（b）所示，因此材料导电性能得到提高。这种掺杂五价元素的杂质半导体主要靠自由电子导电，故自由电子为多数载流子，简称多子；磷原子提供电子，故称为施主原子；这样的杂质半导体为 N 型半导体。同理，如向硅本征半导体中掺入少量三价元素硼 B 后，硼原子外层三个电子只能与相邻三个硅原子的价电子形成共价键，剩下一个硅原子未形成共价键留下空穴，容易吸引电子，如图 7.1（c）所示。这种掺杂三价元素的杂质半导体主要靠空穴导电，故空穴为多数载流子，自由电子则成为少数载流子，简称少子；硼原子提供空穴，故称为受主原子；这样的杂质半导体称为 P型半导体。掺杂浓度越高，多子浓度就越高，导电性能也越强，故控制掺杂浓度可改变杂质半导体的导电性能。应当注意，无外加电压的情况下杂质半导体中正电荷与负电荷数量相等，是不带电的，呈电中性。

Diodes introduced in this chapter, are a type of widely used semiconductor devices. A diode allows a high current to flow from the anode to cathode but a very low current in the reverse direction, aka unilateral conductivity. Taking advantage of this feature, diodes find themselves in switches, rectifiers, clippers and so on. Diodes can also be used for demodulation, voltage regulation, glowing, etc, due to other features.

This chapter begins with some general concepts of the diode such as the structure of semiconductor materials and the characteristics of a PN junction. Afterwards some models of the diode are introduced for circuit analysis. Finally, some practical circuits including rectifier, clipper, clamping circuit and buck converter are discussed.

7.1 General Concepts

7.1.1 Intrinsic Semiconductor and Doped Semiconductor

A semiconductor is a material whose conductivity is between that of a conductor and that of an insulator. The common semiconductor materials are made up of Group IV elements such as silicon and germanium. The pure crystal aka crystalline of these elements is intrinsic semiconductor, in which each of the four valence electrons of an atom forms a covalent bond with one valence electron of its four neighboring atoms, attaining a stable structure shown in Fig. 7.1 (a). Few excited electrons lead to the poor conductivity of the intrinsic semiconductor at room temperature.

Doped semiconductors are formed by adding a tiny amount of dopants to intrinsic semiconductors. For example, after adding a tiny amount of Group V phosphorus to the intrinsic silicon semiconductor, only four of the five valence electrons of a phosphorus atoms can form covalent bonds with its four neighboring silicon atoms, resulting in one valence electron free and easy to excite at room temperature, shown in Fig. 7.1(b). As a result, the conductivity of the material is improved. The current is mainly contributed by the moving electrons in this kind of pentavalent impurity doped semiconductor, thus electrons are called major carriers and phosphorus donating electrons are called donor. The semiconductor doped with Group V elements is called N-type semiconductor. Similarly, after adding a tiny amount of Group III boron atoms to the intrinsic silicon semiconductor, all the three valence electrons of a boron atom form covalent bonds with three of its four neighboring silicon atom, resulting in one silicon atom and its broken bond, that is, a hole, to attract unbound electrons, shown in Fig. 7.1(c). The current is mainly contributed by the moving holes in this kind of trivalent impurity doped semiconductor, thus the holes are major carriers and the electrons turn to be minor carriers. Boron donating holes is called acceptor. The semiconductor doped with Group III elements is called P-type semiconductor. The more the dopants, the better the conductivity. So controlling the dopant concentration can alter the conductivity of the semiconductor. It should be noticed that, doped semiconductors are uncharged because the quantity of the positive charge and the negative charge are equal, and the electrical neutrality is maintained.

7.1.2 PN 结

在同一衬底上掺杂不同杂质形成 P 型半导体和 N 型半导体，这两种半导体的交界面就形成 PN 结。就像放入水中的方糖会因浓度差而扩散至水中，直到形成均匀的溶液一样，N 型半导体中的自由电子和 P 型半导体中的空穴因浓度差而透过交界面扩散至对方并且复合。但这个扩散运动并不会持续到浓度均匀为止，因为界面附近 N 型半导体因自由电子与空穴复合留下不可移动的正离子和 P 型半导体因空穴与自由电子复合留下不可移动的负离子，形成一个空间电荷区，产生一个内电场，该电场使自由电子和空穴做与扩散运动方向相反的漂移运动，如图 7.2 所示。当扩散运动与漂移运动达到动态平衡时，PN 结没有电流流过。由于空间电荷区中自由电子和空穴数量非常少，故也称耗尽层。

PN 结具有单向导电性。当 PN 结 P 端接正极，N 端接负极时，外电场与内电场方向相反，内电场减弱，扩散运动与漂移运动的动态平衡被破坏，扩散运动强于漂移运动，产生从 P 区至 N 区的正向电流，如图 7.3（a）所示，此时 PN 结导通；相反地，当 P 端接负极，N 端接正极，内电场加强，空间电荷区变宽，漂移运动强于扩散运动，产生从 N 区至 P 区的反向电流，如图 7.3（b）所示。由于参与漂移运动的少子数量非常少，故反向电流非常小，认为 PN 结处于截止状态。但当反向电压大于一定值时反向电流将突然增大，此时 PN 结反向击穿。PN 结的伏安特性如图 7.4 所示。

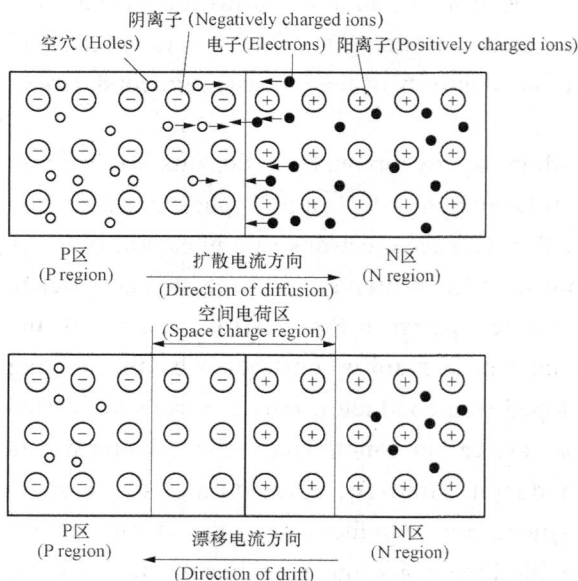

图 7.2　PN 结的形成

Fig. 7.2　The formation of the PN junction

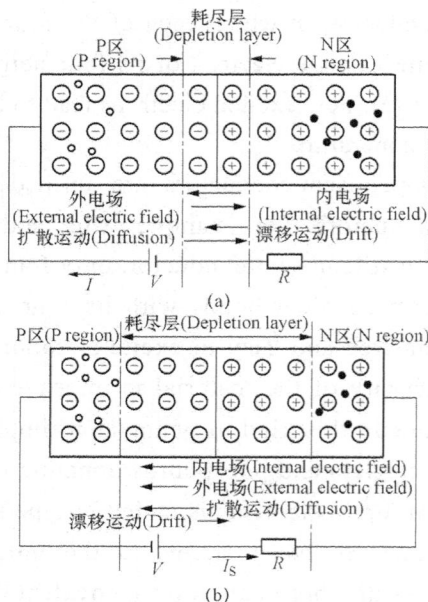

图 7.3　PN 结的单向导电性

（a）PN 结加正向电压时导通；（b）PN 结加反向电压时截止

Fig. 7.3　The unilateral conductivity of the PN junction

(a) When the PN junction is forward biased;

(b) When the PN junction is reverse biased

PN 结流过的电流 I 与端电压 V 的关系为

$$I = I_s(e^{V/V_T} - 1) \tag{7.1}$$

式中，I_s 为反向饱和电流；V_T 为热电压。热电压的表达式为

$$V_T = kT/q \tag{7.2}$$

7.1.2 PN junction

PN junction is the interface between N-type and P-type semiconductors formed by doping a substrate. Just like that the molecules of cube sugar dropped in water diffuse due to the sugar concentration gradient till they form a homogeneous solution, the free electrons from the N-type and holes from the P-type diffuse into the opposite region across the interface and recombine due to the concentration gradient. But the diffusion won't last to the uniform concentration of the electrons and holes. As the electrons diffuse, they leave fixed positively charged ions in the N region near the PN interface; similarly, the holes leave fixed negatively charged ions in the P region. These two regions become charged, and form the space charge region where there is an electric field, forcing the electrons and holes drift in the opposite direction of the diffusion, shown in Fig. 7.2. When the drift and diffusion process reach an equilibrium, the net current flowing through the PN junction is zero. The space charge region is also called depletion layer because of few free electrons and holes there.

The PN junction is unilaterally conductive. When the P side is connected to the anode of a battery and N side to the cathode, the external electrical field opposes the build-in electrical field and alleviates the latter, breaking the equilibrium of the drift and diffusion motion. The diffusion then dominates, causing a current to flow from the P side to the N side, as shown in Fig. 7.3 (a), and the PN junction is on. On the contrary, when the P side is connected to the cathode of a battery while N side to the anode, the external electrical field reinforces the build-in electrical field and the space charge region is widened. The drift then dominates, causing a current to flow from the N side to the P side, as shown in Fig. 7.3(b). But the amount of the minor carriers that take part in this process is very tiny, leading to an also tiny reverse current. Thus the PN junction is considered as off-state. However, once the reverse-biased voltage goes beyond a certain level, the reverse current increases suddenly and the PN junction breaks down. The I-V characteristic of a PN junction is shown in Fig. 7.4.

The relationship between the current I flowing through and voltage V across a PN junction is

$$I = I_{S}(e^{V/V_T} - 1) \tag{7.1}$$

where I_S is the saturation current, and V_T is the thermal voltage. The thermal voltage is defined by

$$V_T = kT/q \tag{7.2}$$

where k is the Boltzmann's constant; T is the absolute temperature and q is the magnitude of the electronic charge. V_T is approximately 26mV at the room temperature (T=300K).

式中，k为玻尔兹曼常数；T为热力学温度；q为电子电量。

常温下，即$T = 300\text{K}$时，V_T约为26mV。

从式（7.1）可见，当PN结正接，且$V \gg 0$时，I约等于$I_S e^{V/V_T}$，电流呈指数变化；当PN结反接，且$V_{(BR)} \ll V \ll 0$时，e^{V/V_T}约为0，I约等于$-I_S$，反向电流基本不变。当$V \ll V_{(BR)}$时，PN结被击穿，反向电流急剧上升。

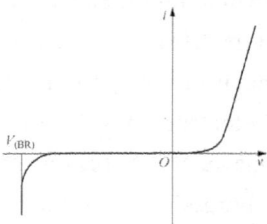

PN结还可看作电容。这个电容由两个部分组成，分别为势垒电容和扩散电容。当外加电压变化时耗尽层宽度发生变化，空间电荷区电荷量随之改变，产生与电容相似的效应，这就是势垒电容；当外加正向电压发生变化时，扩散运动强度也发生变

图 7.4　PN 结伏安特性曲线

Fig. 7.4　The I-V characteristics of the PN junction

化，也产生与电容相似的效应，这就是扩散电容。

7.1.3　二极管

封装好的 PN 结就成为二极管，符号如图 7.5 所示（两根引线分别称为阳极和阴极）。实际测试得到的二极管伏安特性曲线如图7.6 所示。由图可知，当正向电压大于一定值时，二极管电流才开始呈指数增长。这个电压称为开启电压。不同材料的二极管开启电压不同。硅二极管的开启电压约为 0.7V，锗二极管的则约为 0.2V。

图 7.5　二极管符号

Fig. 7.5　The symbol of the diode

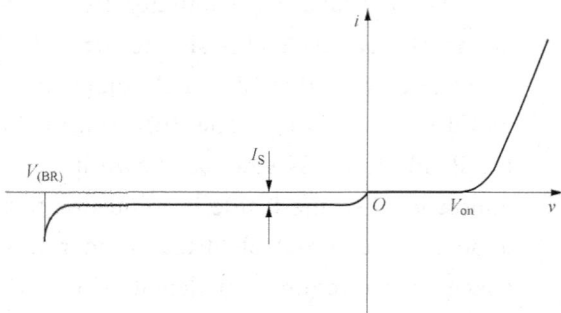

图 7.6　二极管的伏安特性曲线

Fig. 7.6　The I-V characteristic of the diode

7.2　理想二极管的伏安特性与电路模型

理想二极管的开启电压为零，反向电流为零，伏安特性曲线如图7.7 所示。理想二极管导通时电阻为零，相当于开关闭合；截止时电阻无穷大，相当于开关断开，如图7.8 如示，用一个空心二极管符号表示。在精度要求不高时，使用理想二极管模型会便于分析电路。

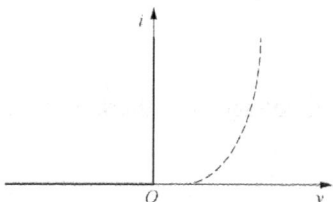

图 7.7　理想二极管伏安特性曲线

Fig. 7.7　The I-V characteristics of the ideal diode

图 7.8　理想二极管符号

Fig. 7.8　The symbol of the ideal diode

Equation (7.1) shows that, when the PN junction is forward biased and $V \gg 0$, I approximately equals to $I_{\mathrm{s}}e^{V/V_{\mathrm{T}}}$ and the current changes with the voltage exponentially; when the PN junction is reverse-biased and $V_{(\mathrm{BR})} \ll V \ll 0$, $e^{V/V_{\mathrm{T}}}$ is nearly 0 and I approximately equals to $-I_{\mathrm{s}}$, showing that the reverse current remains unchanged. When $V \ll V_{(\mathrm{BR})}$, the diode breaks down and the reverse current increases dramatically.

The PN junction can also be regarded as a capacitor. This capacitor is made up of two components, depletion capacitance and diffusion capacitance, respectively. When the voltage across the PN junction varies, the width of the depletion layer as well as the charge stored in the space charge region also varies, acting like a capacitor, named as depletion capacitance. When the voltage of the forward bias changes, the amount of the minor carriers sweeping through the depletion layer to the opposite region due to the diffusion also changes, acting like a capacitor as well, named as diffusion capacitance.

7.1.3 Diode

The PN junction after packing becomes a diode, whose symbol is shown in Fig. 7.5. The actual I-V characteristic of a diode is shown in Fig. 7.6. It shows that the current does not increase until the forward bias voltage goes beyond a certain value. The voltage needed to turn on the diode is called threshold voltage, and it varies among diodes of different materials. The threshold voltage of the silicon diode is about 0.7V, while that of the germanium diode is about 0.2V.

7.2 I–V Characteristics and Model of Ideal Diode

The threshold voltage and reverse current of an ideal diode are both zero, with I-V characteristic shown in Fig. 7.7. The ideal diode acts like a short circuit with zero resistance when forward biased and an open circuit with infinite resistance when cut off. The ideal diode is symbolized by an empty triangle as shown in Fig. 7.8. It is convenient to analyze the circuit with the ideal diode model when the precision does not really matter.

More facts about diodes

The most common function of a diode is to allow an electric current to pass in one direction (called the diode's forward direction), while blocking current in the opposite direction (the reverse direction). However, diodes can have more complicated behavior than this simple on-off action, because of their nonlinear current-voltage characteristics. Semiconductor diodes begin conducting electricity only if a certain threshold voltage is present in the forward direction. The voltage drop across a forward-biased diode varies only a little with the current, and is a function of temperature; this effect can be used as a temperature sensor or as a voltage reference.

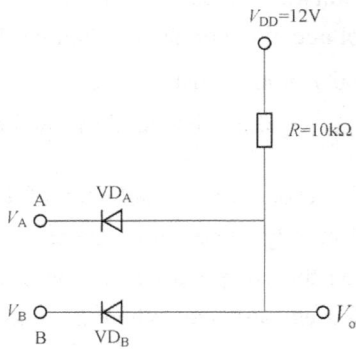

图 7.9 例 7.1 电路图

Fig.7.9 Circuit diagram for Example 7.1

【例 7.1】 如图 7.9，试使用理想二极管模型分析输出电压 V_o，当

a）V_A=0V，V_B=0V；

b）V_A=0V，V_B=5V；

c）V_A=5V，V_B=0V；

d）V_A=5V，V_B=5V。

解 a）当 V_A 和 V_B 为 0V 时，VD$_A$ 和 VD$_B$ 导通，相当于开关闭合短路，如图 7.10（a）所示，此时 V_o 为 0V；

b）当 V_A 为 0V，V_B 为 5V 时，VD$_A$ 导通相当于开关闭合，VD$_B$ 截止相当于开关断开，如图 7.10（b）所示，此时 V_o 为 0V；

c）同理当 V_A 为 5V，V_B 为 0V 时，VD$_A$ 截止，VD$_B$ 导通，如图 7.10（c）所示，此时 V_o 为 0V；

d）当 V_A 和 V_B 为 5V 时，VD$_A$ 和 VD$_B$ 导通，如图 7.10（d）所示，此时 V_o 为 5V。
结果整理在图 7.10（e）中。

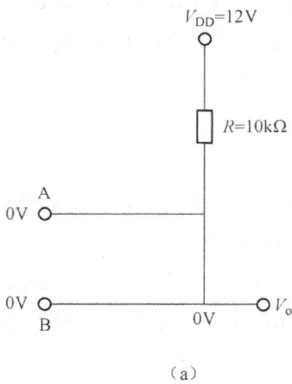

（a）

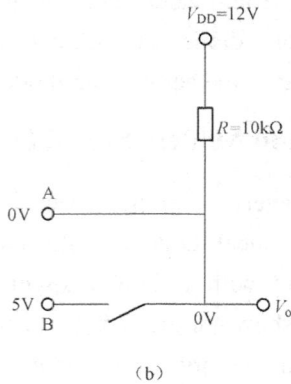

（b）

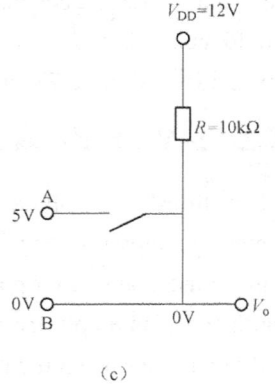

（c）

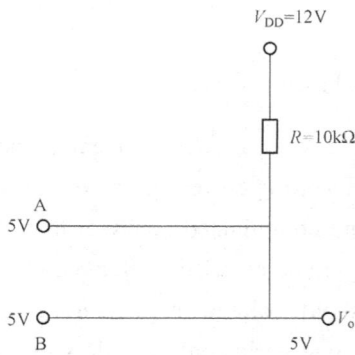

（d）

		理想二极管 (An ideal diode)		输出电压 (Output voltage) (V)
V_A(V)	V_B(V)	VD$_A$	VD$_B$	
0	0	饱和 (Saturated)	饱和 (Saturated)	0
0	5	饱和 (Saturated)	截止(Cut-off)	0
5	0	截止(Cut-off)	饱和 (Saturated)	0
5	5	饱和 (Saturated)	饱和 (Saturated)	5

（e）

图 7.10 例 7.1 求解

Fig. 7.10 Solution to Example 7.1

Example 7.1: Try to find out the output voltage V_o of the circuit shown in Fig. 7.9 with the model of ideal diode when

　　a）V_A=0V, V_B=0V;

　　b）V_A=0V, V_B=5V;

　　c）V_A=5V, V_B=0V;

　　d）V_A=5V, V_B=5V.

Solution:

　　a）When both V_A and V_B are 0V, VD_A and VD_B are on, just like short circuits as shown in Fig. 7.10(a). Thus V_o is 0V;

　　b）When V_A is 0V and V_B is 5V, VD_A is short and VD_B is open, as shown in Fig. 7.10(b). Thus V_o is 0V;

　　c）Similarly, when V_A is 5V and V_B is 0V, VD_A is cut off and VD_B is on, as shown in Fig. 7.10(c). Thus V_o is 0V;

　　d）When both V_A and V_B are 5V, both VD_A and VD_B are on, as shown in Fig. 7.10(d). Thus V_o is 5V.

　　The results are summarized in Fig. 7.10(e).

More facts about diodes

Thermionic (vacuum tube) diodes and solid state (semiconductor) diodes were developed separately, at approximately the same time, in the early 1900s. They were used as radio receiver detectors, and until the 1950s, vacuum tube diodes were used more frequently because the early point-contact type semiconductor diodes were less stable.

In 1873, Frederick Guthrie discovered the basic principle of operation of vacuum tube diodes. Guthrie discovered that a positively charged electroscope could be discharged by bringing a grounded piece of white-hot metal close to it (but not actually touching it). The same did not apply to a negatively charged electroscope, indicating that the current flow was only possible in one direction. Thomas Edison independently rediscovered the principle in 1880, and devised a circuit which produced one-way current, the phenomenon of which was later on referred Edison effect by a British scientist. About 20 years later, John Ambrose Fleming realized that the Edison effect could be used as a precision radio detector.

In 1874 German scientist Karl Ferdinand Braun discovered the unilateral conduction of crystals, and crystal was used for detecting radio waves later on. The crystal detector was developed into a practical device for wireless telegraphy. Early crystal detectors suffered from some drawbacks and were superseded by vacuum tube diodes by the 1920s. But in the 1950s, after high purity semiconductor materials became available, the crystal detector returned to dominant use with the advent of inexpensive fixed-germanium diodes.

7.3 齐纳二极管的电路模型

从前面二极管的伏安特性曲线可看到，当外加反向电压大于一定值时反向电流将突然增大，这时二极管反向击穿，如图 7.11 所示。反向击穿时即使反向电流变化很大，反向电压也几乎不变，内阻很小，具有稳压特性。齐纳二极管就是工作在这种状态下具有稳压特性的二极管，又称为稳压管，其符号如图 7.12 所示。正向导通和反向未击穿时，齐纳二极管与一般二极管无异；反向击穿后，齐纳二极管相当于一个电压源和一个电阻串联，其模型如图 7.13 所示。图中 V_Z 为齐纳二极管的反向击穿后稳定工作时的电压，称为稳定电压；r_Z 为反向击穿后的内阻。

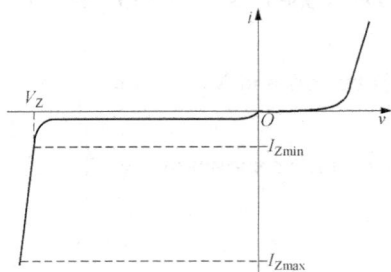

齐纳二极管须工作在一定的电流范围内，电流过小时稳压效果不好，电流超过额定值则因温度过高而损坏。这个最小的电流称为最小稳定电流，用 I_{Zmin} 表示，最大的电流称为最大稳定电流，用 I_{Zmax} 表示。故齐纳二极管常串联一个限流电阻使用。

图 7.11　稳压二极管反向击穿的伏安特性

Fig. 7.11　The I-V characteristics of the diode in the breakdown region

图 7.12　齐纳二极管符号

Fig. 7.12　The symbol of the Zener diode

【例 7.2】 如图 7.14 所示，在稳压电路中，输入电压 V_i =8 V，稳压管的稳定电压 V_Z =4 V，最小稳定电流 I_{Zmin} =10 mA，最大稳定电流 I_{Zmax} =20 mA，负载电阻 R_L =400 Ω，求限流电阻 R 的取值范围。

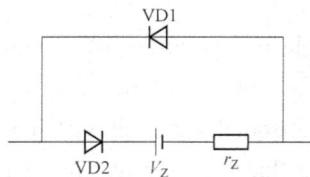

图 7.13　齐纳二极管模型

Fig. 7.13　The model of the Zener diode

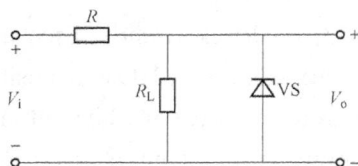

图 7.14　例 7.2 电路图

Fig. 7.14　Circuit diagram for Example 7.2

解　R 上的电流等于稳压二极管电流 I_Z 和负载电阻电流 I_L 之和，而

$$I_L = V_Z / R_L =(4/400)A=10 \text{ mA}$$
$$I_Z=10 \sim 20 \text{ mA}$$

所以

$$I_R =20 \sim 30 \text{ mA}$$

因为

$$V_R = V_i - V_Z =(8-4)V=4 \text{ V}$$

7.3   Model of Zener Diode

The I-V characteristic of the diode in the previous section shows that the reverse current increases suddenly and the diode breaks down when the reverse-bias voltage goes beyond a certain level, as shown in Fig. 7.11. In the breakdown region, the reverse-bias voltage remains nearly unchanged though the reverse current changes a lot. The Zener diode is such a type of diode that works in the breakdown region, and is usually used as a voltage regulator. Its symbol is shown in Fig. 7.12. When it is forward biased or unbroken down, the Zener diode behaves like a normal diode. But when it breaks down, the Zener diode acts like a voltage source and a resistor connected in series, whose model is shown in Fig. 7.13, where V_Z, called Zener voltage, denotes the voltage beyond which the Zener diode starts to work steadily and r_Z denotes the resistance in the breakdown region.

The Zener diode only works as expected within a certain current range. It does not work properly when the current is too small; otherwise, it would be damaged due to high temperature when the current was too large. The lower limit of the current range is denoted by I_{Zmin}, while the upper limit is denoted by I_{Zmax}. Therefore, we usually put a current-limiting resistor in series with the Zener diode.

Example 7.2: Determine the resistance range of the current-limiting resistor R in the voltage regulation circuit as shown in Fig. 7.14, where the input voltage V_i=8 V, the Zener voltage V_Z=4 V, the minimum allowable current I_{Zmin}=10 mA, the maximum allowable current I_{Zmax}=20 mA and the load resistor R_L=400 Ω.

Solution:

The current flowing through R is the sum of I_Z through the Zener diode and I_L through the load. Since

$$I_L = V_Z / R_L = 4/400 = 10 \text{ mA}$$

$$I_Z = 10 \sim 20 \text{ mA}$$

we have

$$I_R = 20 \sim 30 \text{ mA}$$

Since

$$V_R = V_i - V_Z = 8 - 4 = 4 \text{ V}$$

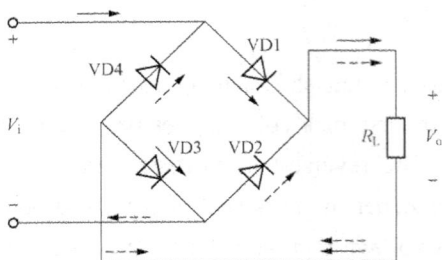

图 7.15 二极管全波整流电路

Fig. 7.15 A diode single-phase full-wave rectifier

所以

$$R_{max} = V_R / I_{R\,min} = (4/0.02)\ \Omega = 200\ \Omega$$

$$R_{min} = V_R / I_{R\,max} = (4/0.03)\ \Omega = 133\ \Omega$$

即限流电阻 R 的取值范围是 $133 \sim 200\ \Omega$。

✒ 讨论： 整流、限幅、钳位、斩波电路分析

1. 整流电路

单相全波整流电路如图 7.15 所示，输入峰值 V_i 为 50V 的正弦波电压，请同学们：

（1）分析电路如何实现整流功能；

（2）计算电路输出电压平均值和输出电流平均值；

（3）假设输入电压有 10%的波动，应选择能承受多大反向电压的二极管。

分析：

当 V_i 为正半周时，VD1 和 VD3 导通，VD2 和 VD4 截止，电流从电阻 R 上端流向下端，电流方向如图 7.15 中实线箭头所示；当 V_i 为负半周时，VD1 和 VD3 截止，VD2 和 VD4 导通，电流也是从电阻 R 上端流向下端，电流方向如图 7.15 虚线箭头所示。输出电压与电流的波形如图 7.16 所示。因此该电路能实现整流功能。

在一个周期内，有

$$V_{oavg} = \frac{1}{2\pi}\int_0^{2\pi} |V_i \sin \omega t|\, \mathrm{d}(\omega t) = \frac{2}{\pi} V_i$$

$$I_{oavg} = V_{oavg} / R$$

应选择能承受最大反向电压为 $V_{(BR)} = 1.1 \times 50 = 55$ V 的二极管。

2. 限幅电路

限幅电路通常用于限制电路中某一点的信号幅值的大小。当信号的幅值没有达到限定值时，限幅电路不工作；但当信号幅值超过限值范围时，限幅电路工作，限制信号的幅值再增大或减小。可以利用二极管的单向导电以及导通压降基本为定值的特性来设计限幅电路。如图 7.17 所示，$V_i = 10\sin \omega t$ V，设二极管导通压降 $V_D = 0.7$V，试分析输入电压与输出电压的关系，画出 V_i 和 V_o 的波形。

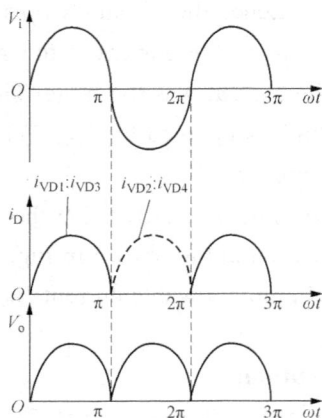

图 7.16 二极管全波整流电路的波形

Fig. 7.16 The waveforms of the diode single-phase full-wave rectifier

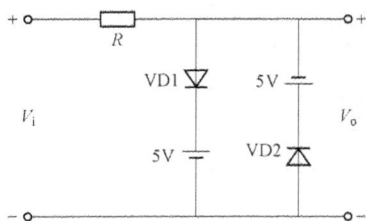

图 7.17 限幅电路

Fig. 7.17 Clipper circuit

分析：

（1）当 $-5.7\ \text{V} \leqslant V_i \leqslant 5.7\ \text{V}$，二极管 VD1、VD2 截止，$V_o = V_i$；

（2）当 $V_i \leqslant -5.7\ \text{V}$，二极管 VD1 截止、VD2 导通，$V_o = -5.7\text{V}$；

（3）当 $V_i \geqslant 5.7\ \text{V}$，二极管 VD1 导通、VD2 截止，$V_o = 5.7\text{V}$。

we can get

$$R_{max} = V_R / I_{R\,min} = (4/0.02)\ \Omega = 200\ \Omega$$

$$R_{min} = V_R / I_{Rmax} = (4/0.03)\ \Omega = 133\ \Omega$$

Thus the current-limiting resistor is in a range of 133 Ω to 200 Ω.

Discussion:　Analysis of rectifier, clipper, clamping circuit and buck converter

1.　Rectifier

A single-phase full-wave rectifier is shown in Fig. 7.15, and the input is a sine wave with peak amplitude of 50V. Please

(1) Explain the operation of the rectifier;

(2) Determine the mean value of the output voltage and current;

(3) Assume the input voltage fluctuates 10%, please select the optimum diode which can sustain the maximum reverse-bias voltage in the circuit.

Analysis:

When V_i goes positive, VD1, VD3 are on, and VD2, VD4 are off. Thus the current through R flows from top to bottom, as shown by solid line in Fig. 7.15. When V_i goes negative, VD1, VD3 are off, and VD2, VD4 are on. Thus the current through R also flows from top to bottom, as shown by dashed line in Fig. 7.15. The waveforms of the output voltage and current are shown in Fig. 7.16 . Thus the rectification is realized.

Within one cycle, there is

$$V_{oavg} = \frac{1}{2\pi}\int_0^{2\pi} |V_i \sin \omega t|\, d(\omega t) = \frac{2}{\pi} V_i$$

$$I_{oavg} = V_{oavg} / R$$

If the input voltage varies in the range of 10%, the maximum reverse-bias voltage of the diode in the circuit is 1.1×50=55 V. The breakdown voltage of selected diode is $V_{(BR)} > 55$ V.

2.　Clipper Circuit

Clipper circuit is usually used to limit the amplitude of signal at some node. When the amplitude is within the specified range, the clipper is inactive; while the signal amplitude is over the specified range, the clipper starts to prevent it from further increasing or decreasing. The characteristics of diode including the unilateral conduction and nearly fixed turn-on voltage are explored to design the clipper. As shown in Fig. 7.17, $V_i=10\sin \omega t$ V, with the assumption of V_D =0.7 V, please determine and plot the relationships of V_i and V_o.

Analysis:

（1）when -5.7 V$\leqslant V_i \leqslant 5.7$ V, both VD1 and VD2 are turned off, and $V_o = V_i$;

（2）when $V_i \leqslant -5.7$ V, VD1 is off while VD2 is on, and $V_o = -5.7$ V;

（3）when $V_i \geqslant 5.7$ V, VD1 is on while VD2 is off, and $V_o = 5.7$ V;

V_o 与 V_i 的波形图如图 7.18 所示。

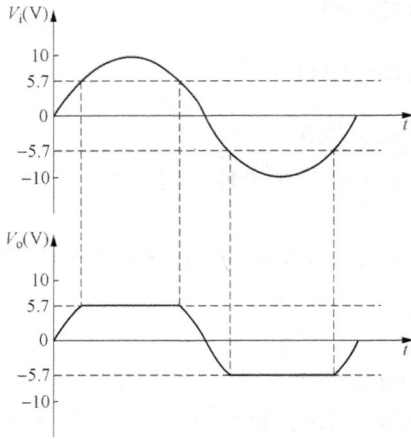

图 7.18 限幅电路的波形
Fig. 7.18 The waveforms of a clipper circuit

3. 箝位电路

箝位电路也称直流恢复器，它可将波形的上端或下端固定至一个直流电压上。图 7.19 为箝位电路图，假设输入为一周期为 T、峰值为 V 的方波，其中二极管 VD 正向导通电阻为 R_D，且有时间常数 $R_D C \ll T \ll RC$，$V_{REF} \ll V$，分析输出电压的波形。

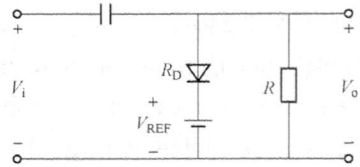

图 7.19 箝位电路
Fig. 7.19 A clamping circuit

分析：

当 V_i 上升后，二极管 VD 导通，内阻 R_D 远小于 R，电阻 R 支路相当于开路，此时电路相当于图 7.20（a），并且由于时间常数 $R_D C \ll T$，电容迅速充电后输出 V_o 很快就达到 V_{REF}，电容两端电压 V_C 大小为 $(V - V_{REF})$；当 V_i 突然下降后，二极管 VD 截止，相当于开路，此时电路相当于图 7.20（b），但由于时间常数 $RC \gg T$，电容放电缓慢，两端的电压基本保持不变，故输出 V_o 基本保持在 $0 - V_C = 0 - (V - V_{REF}) = V_{REF} - V$。输出波形如图 7.20（c）所示，可见该电路将方波上端移至 V_{REF} 处。

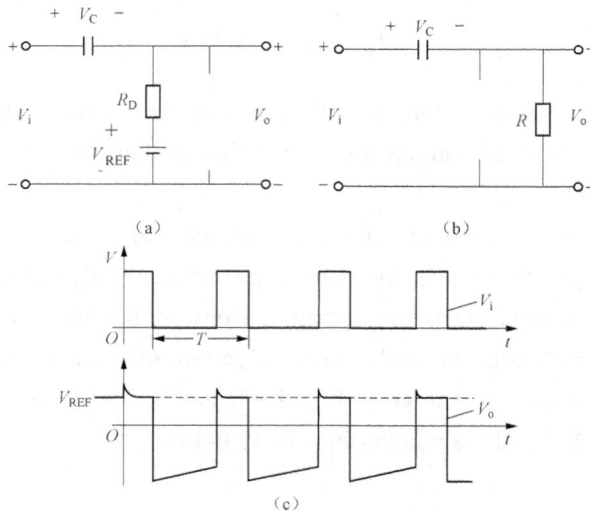

图 7.20 箝位电路分析
（a）输入电压升高时的等效电路；（b）输入电压降低时的等效电路；（c）电路的波形
Fig. 7.20 Analysis of the clamping circuit
(a) The equivalent circuit when the input goes up; (b) The equivalent circuit when the input goes down; (c) The waveforms

The waveforms of V_o and V_i are plotted in Fig. 7.18.

3. Clamping circuit

A clamping circuit, aka DC voltage restorer, can fix either the positive or the negative peak at a DC value. A clamping circuit is shown in Fig. 7.19, where the input is a square wave with a period T and peak amplitude V; R_D is the on-resistance of the diode VD, and the time constants $R_D C \ll T \ll RC$, $V_{REF} \ll V$. Please analyze the output voltage.

Analysis:

When V_i goes positive, the diode VD is on and its on-resistance R_D is much less than R. The branch R becomes an open circuit and the circuit should look like Fig. 7.20 (a). Because the time constant $R_D C \ll T$, the capacitor completes charging quickly and the output obtains V_{REF}. The capacitor voltage V_C is $V - V_{REF}$. When V_i goes negative, the diode VD is off and the branch of VD becomes an open circuit as shown in Fig. 7.20(b). But the time constant $RC \gg T$, the capacitor discharges slowly and its voltage remains nearly unchanged. Then the output V_o keeps $0 - V_C = 0 - (V - V_{REF}) = V_{REF} - V$. The waveform of the output is shown in Fig. 7.20(c), where the positive peak of the square wave is moved to V_{REF}.

More facts about diodes

A semiconductor diode's current-voltage characteristic can be tailored by selecting the semiconductor materials and the doping impurities introduced into the materials during manufacture. These techniques are used to create special-purpose diodes that perform many different functions. For example, diodes are used to regulate voltage (Zener diodes), to protect circuits from high voltage surges (avalanche diodes), to electronically tune radio and TV receivers (varactor diodes), to generate radio-frequency oscillations (tunnel diodes, Gunn diodes, IMPATT diodes), and to produce light (light-emitting diodes). Tunnel, Gunn and IMPATT diodes exhibit negative resistance, which is useful in microwave and switching circuits.

4. 斩波电路

对直流电源进行降压，最简便的方法应该是使用如 7805 之类的线性调节器，但同时这也是十分浪费电能的做法。斩波电路则是一种高效的直流电压转换电路。图 7.21 所示是一个降压斩波电路的原理电路图。假设图中电容 C 足够大，请分析：

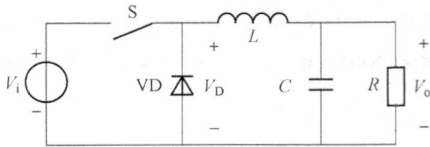

图 7.21　降压斩波电路

Fig .7.21　A voltage step down buck converter

（1）S 处于导通和断开状态时电流流向；

（2）若 S 由一周期为 T、占空比为 D 的方波信号控制通断状态，分析斩波电路为何能实现降压功能，以及输出电压 V_o 与输入电压 V_i 的关系。

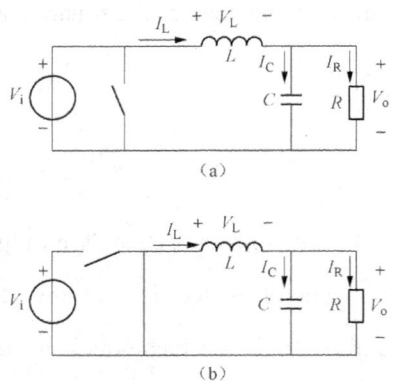

分析：

（1）当 S 处于闭合状态时，二极管 VD 处于截止状态，相当于开路，电流从电源正极经过电感 L 流过负载电阻 R 和滤波电容 C 后回到电源负极，如图 7.22（a）所示；当 S 处于断开状态时，电感储能释放，此时二极管 VD 处于导通状态，电流流过负载电阻 R 和滤波电容后经过二极管 VD 流回电感，如图 7.22（b）所示。

（2）先进行定性分析，假设电路已进入一个稳定的循环状态，即一个控制信号方波周期开始和结束时的电路状态相同。当 S 导通时电源 V_i 为负载供电，电流 I_L 线性增大，使电感 L 两端产生固定的电压降，其中一部分大小不变的电流 I_R 流过负载电阻 R，另一部分大小变化的电流 I_C 则流过电容 C（之前假设电容足够大，以使电流流过时其两端电压不变），输出电压 V_o 不变。当 S 断开时电感 L 储能释放，可视作临时的电流源，经二极管 VD 维持电流，电流 I_L 线性减小（此处假设一个周期完成时电流仍未减小至 0），使电感 L 两端产生与之前方向相反且大小与输出电压相同的电压降 $V_L=-V_o$，同时电容提供电流实现滤波以保持流过负载电阻 R 的电流 I_R 不变（电容足够大，两端电压不变），输出电压 V_o 保持不变。控制信号、二极管两端电压 V_D、电感两端电压 V_L 和电感电流 I_L 以及输出电压 V_o 的波形如图 7.23 所示。

图 7.22　降压斩波电路的等效电路

（a）当 S 导通时；（b）当 S 断开时

Fig .7.22　Equivalent circuits of the voltage step down buck converter

(a) When S is on; (b) When S is off

（3）再进行定量分析，由于电感两端电压 $V_L = L\mathrm{d}I/\mathrm{d}t$，在 S 闭合时，$V_L = V_i - V_o$；在断开时，$V_L = -V_o$，得

$$\mathrm{d}I_{\mathrm{ON}} = \int_0^{DT} \frac{V_L}{L}\mathrm{d}t = \frac{V_i - V_o}{L}DT$$

和

$$\mathrm{d}I_{\mathrm{OFF}} = \int_{DT}^{T} \frac{V_L}{L}\mathrm{d}t = \frac{-V_o}{L}(1-D)T$$

4. Buck converter

The simplest way to reduce the voltage of a DC supply is to use a linear regulator such as 7805, but it is less energy efficient. Buck converters, on the other hand, can be an efficient solution. The basic circuit diagram of a voltage step down buck converter is shown in Fig. 7.21, where the capacitor C is assumed big enough. Please

(1) Find out the current direction when switch S is on and off;

(2) If S is controlled by a square wave whose period is T and duty is D, please explain how the voltage step-down is realized and the relationship between the output voltage V_o and the input voltage V_i.

Analysis:

(1) When S is closed, diode VD is cut off as an open circuit, and the current flows from the positive polarity of the voltage source through inductor L, load resistor R and filter capacitor C, then backs to the negative polarity of the voltage source as shown in Fig. 7.22 (a). When S is open, the inductor releases its energy. The diode is on and the current flows from the inductor, then through the resistor and the capacitor, through the diode, and finally backs to the inductor as shown in Fig. 7.22(b).

(2) Firstly, we do the qualitative analysis. If we assume that the circuit operates in a steady state, the state of the circuit at the beginning of one cycle is the same as the one at the end of the same cycle. When S is on, the source V_i supplies the circuit and current I_L increases linearly, resulting in a constant voltage drop across the inductor. The constant part of this current, I_R, flows through the resistor, and the variable part I_C flows through the capacitor (the capacitor is assumed so big that the capacitor voltage V_o remains nearly the same as the current flows through the capacitor). The output voltage V_o is a constant. When S is open, the inductor, used as a temporarily source, releases the energy it stores and maintains the current through the diode, and current I_L decreases linearly (we assume that the current does not fall to zero at the end of the cycle), producing a voltage drop V_L $=-V_o$ that is as large as the output voltage but in the reverse polarity. At the same time, the capacitor provides part of the current to keep I_R constant (the capacitor is big enough and the voltage across it remains unchanged). The output voltage, V_o, still maintains as a constant. The waveforms of the control signal, the diode voltage V_D, the inductor voltage V_L, the current I_L and the output voltage V_o are shown in Fig. 7.23.

(3) Now we do the quantitative analysis. The voltage across a inductor is defined by

$$V_L = L\mathrm{d}I / \mathrm{d}t$$

When S is on,

$$V_L = V_i - V_o$$

and when S is off,

$$V_L = -V_o$$

we can get

$$\mathrm{d}I_{ON} = \int_0^{DT} \frac{V_L}{L}\mathrm{d}t = \frac{V_i - V_o}{L}DT$$

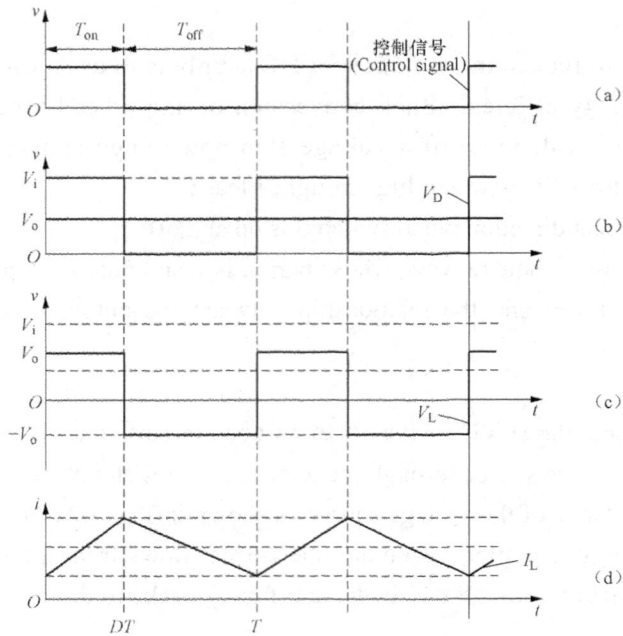

图 7.23 降压斩波电路的波形图

Fig. 7.23 The waveforms of the voltage step down buck converter

由于电路处于稳定状态，一周期后电感电流不变，即 $\mathrm{d}I_{\mathrm{ON}} + \mathrm{d}I_{\mathrm{OFF}} = 0$，可得

$$V_{\mathrm{o}} = DV_{\mathrm{i}} = \frac{t_{\mathrm{ON}}}{T}V_{\mathrm{i}}$$

以上假设一个周期完成时电感电流仍未减小至 0。若电感电流减小至 0，即周期开始和结束时电感电流均为 0，情况比较复杂，输出电压 V_{o} 还与电感 L 和负载电流 I_{o} 有关，为

$$V_{\mathrm{o}} = V_{\mathrm{i}} \frac{1}{\dfrac{2LI_{\mathrm{o}}}{D^2 V_{\mathrm{i}} T} + 1}$$

and

$$dI_{OFF} = \int_{DT}^{T} \frac{V_L}{L} dt = \frac{-V_o}{L}(1-D)T$$

Since the circuit operate S in a steady state, the current of the inductor remains unchanged after one cycle, that is $dI_{ON} + dI_{OFF} = 0$, and we get

$$V_o = DV_i = \frac{t_{ON}}{T} V_i$$

We assume that the current does not fall to zero at the end of a cycle in the analysis above. It becomes more complicated if that current falls to zero, or in other words, the currents of the inductor are zero both at the beginning and at the end of a cycle. Under this condition, output voltage V_o is related to the inductance of the inductor, L, and load current I_o, and V_o is written by

$$V_o = V_i \frac{1}{\dfrac{2LI_o}{D^2 V_i T} + 1}$$

▶ Problems

P7.1   Find the values of V_o in the circuits shown in Fig. P7.1 and explain the bias condition of the diodes, reverse-bias or forward-bias?

P7.2   As shown in Fig. P7.2, $V_{DD}=3$ V, $R=4.7$ kΩ, please determine the V_o. VD1 and VD2 are assumed to be ideal diodes.

(1) $V_1=0$ V, $V_2=0$ V;

(2) $V_1=0$ V, $V_2=3$ V;

(3) $V_1=3$ V, $V_2=3$ V.

P7.3 As shown in Fig. P7.3, $V_i= 10\sin \omega t$ V, if the turn-on voltage is ignored, please plot the V_i and V_o.

P7.4 As shown in Fig. P7.4, $V_i =10$ V, Zener voltage $V_z=6$ V, $I_{Zmin}=5$ mA, the maximum power consumption the Zener diode can sustain is $P_Z=150$ mW. Determine the resistance limits of R.

P7.5 As shown in Fig. P7.5, in the clipper circuit, the input is $V_i=10\sin \omega t$ V. Plot the V_o and the transform characteristics of the circuit. VD1, VD2 are assumed to be ideal diodes.

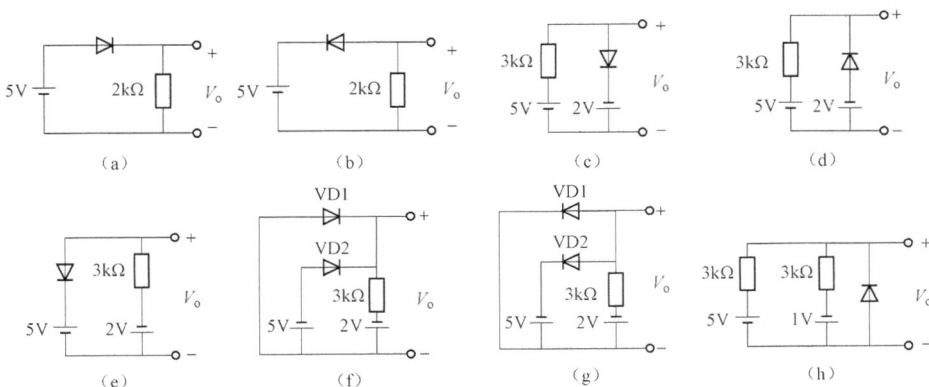

Fig. P7.1

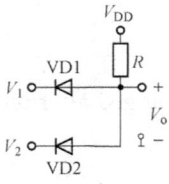

Fig. P7.2

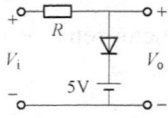

Fig. P7.3

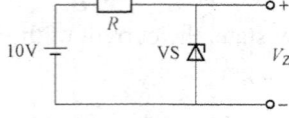

Fig. P7.4

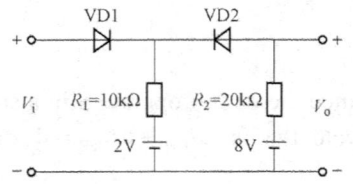

Fig. P7.5

双极型三极管
Bipolar Junction Transistors

双极型半导体三极管是由两个背对背的 PN 结构成的电流控制型器件。由于器件内部依靠两种不同极性的载流子参与导电，故得名为"双极"。通常所说的三极管就是指双极型半导体三极管。

本章将介绍三极管的工作原理与伏安特性，然后分别对三极管放大电路进行大信号和小信号分析，并在此基础上讨论几种常见的三极管放大电路的实际应用。

8.1 三极管的基本概念与伏安特性

8.1.1 三极管的基本工作原理

三极管通过在一块硅片上形成两个背靠背的 PN 结制成，其结构示意图如图 8.1 所示。它有发射极、基极和集电极三个电极，分别以字母 e、b、c 表示。其中图 8.1（a）为 NPN 型三极管的结构图，它以 N 型衬底作为集电极，通过在衬底上扩散出 P 型材料形成基区，再在基区上注入高掺杂浓度的 N 型材料形成发射极。PNP 型三极管则通过在 P 型衬底上扩散 N 型材料形成基区，再注入 P 型材料制成发射极，其结构如图 8.1（b）所示。三极管均由三层材料构成，其中发射极的材料掺杂浓度很高，在正向偏置时能形成大量的扩散载流子；基区材料非常薄且掺杂浓度比发射极低几个数量级，使得只有少数由发射区扩散过来的载流子会被复合；集电极的面积很大，在反向偏置时利于收集在基区没被复合的载流子。

三极管电路符号如图 8.2 所示，其中箭头的方向代表发射极电流的方向。图 8.2（a）为 NPN 型三极管符号，箭头向外表示电流从基极流向发射极；图 8.2（b）为 PNP 型三极管符号，箭头向内表示电流从发射极流向基极。本章将以 NPN 型三极管为例讨论三极管的基本工作原理及伏安特性，而对 PNP 型三极管的分析则留给读者自己完成。

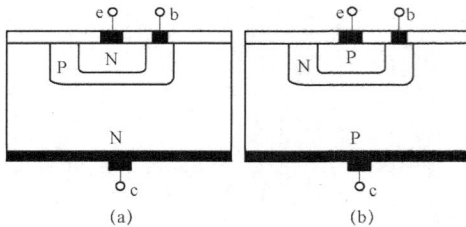

图 8.1 三极管结构示意图

（a）NPN 型三极管；（b）PNP 型三极管

Fig. 8.1 An illustration of BJT's physical structure

(a) An NPN BJT; (b) A PNP BJT

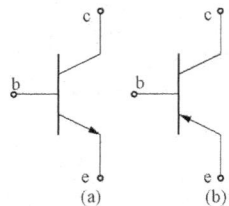

图 8.2 三极管电路图

（a）NPN 型三极管；（b）PNP 型三极管

Fig. 8.2 Circuit symbol of transistors

(a) An NPN BJT; (b) A PNP BJT

8.1.2 工作在放大区

考虑三极管如图 8.3 所示的连接情况，即三极管发射极正向偏置，集电极反向偏置的状态。此时三极管工作于放大区，基极电流微小的变化将会引起集电极电流很大的变化，三极管具有放大的作用。以下将以 NPN 型三极管为例详细分析此时三极管内部载流子的流动情况。

1. 发射区多数载流子的扩散运动形成发射极电流

因为发射结加正向电压，发射区的多数载流子电子扩散到基区，同时基区中的多数载流子空穴扩散到发射区，形成发射极电流。发射极电流即为正向偏置下 PN 结的扩散电流。但在一般情况下，发射区的掺杂浓度很高，提供了大量的电子，而基区的掺杂浓度比发射区的掺

Bipolar junction transistors are a type of current-controlled devices, consisting of two back-to-back PN junctions. The name 'bipolar' comes from the fact that both electrons and holes are carriers. Generally, the term transistor refers to bipolar junction transistor, abbreviated as BJT.

This chapter begins with some general concepts such as the principle and I-V characteristics of the transistor. Then we do the large-signal and small-signal analysis of transistor amplifiers. Finally, a few types of practical amplifiers are discussed.

8.1   General Concepts and I–V Characteristics

8.1.1   General concepts of transistors

A transistor is formed by two back-to-back PN junctions doped on the same substrate and its structure is shown in Fig. 8.1. It has three terminals, labeled with emitter (e), base (b), and collector (c) respectively. The structure of NPN transistor is shown in Fig. 8.1(a). The N-type substrate serves as the collector. Then the P-doped layer formed by ion diffusion on the substrate servers as the base, and the emitter is the heavily doped N region formed by ion implantation on the base. Similarly, for a PNP transistor whose structure is shown in Fig. 8.1(b), the P-type substrate serves as the collector. Then the N-doped layer formed by ion diffusion on the substrate servers as the base, and the emitter is the heavily doped P region formed by ion implantation on the base. Among the three regions of a transistor, the emitter is doped so heavily that it can contribute a large number of carriers when the junction is forward-biased. The base is a very thin layer doped several orders of magnitude smaller than the emitter, leading to few carriers diffused from the emitter recombining here. For the collector, its area is comparably large and easy to collect the carriers that manage to escape the recombination in the base. The symbols of transistors are shown in Fig. 8.2, where the arrowhead points the directions of the emitter currents. The symbol of NPN transistor is shown in Fig. 8.2(a) with an outgoing arrowhead denoting the current flows from base to emitter. The symbol of PNP transistor is shown in Fig. 8.2(b) with an ingoing arrowhead denoting the current flows from emitter to base. In this chapter, the basic operation of NPN transistors is discussed, while the analysis of PMOS transistors is handed over to readers.

8.1.2   Operation in active region

Consider the transistor in the configuration shown in Fig 8.3, where the emitter-base junction is forward-biased and the collector-base junction is reverse-biased. The transistor then operates in the active region. A small current change in the base can cause a big change current in the collector and, namely, the transistor amplifies the current variation. We are going to talk about the carriers in detail taking the NPN transistor as example.

1.   Emitter current: the diffusion of the major carriers

Due to the forward-biased emitter-base junction, the major carriers from the emitter, namely electrons, diffuse to the base, and the major carriers from the base, namely holes, diffuse to the emitter simultaneously , forming the emitter current. That is, the emitter current is the diffusion current of a forward-biased PN junction. But generally the emitter is heavily doped and provides tons of electrons, while the base is doped several orders of magnitude smaller than the emitter, thus

杂浓度低 2～3 个数量级，故可忽略空穴的扩散电流。在近似分析时，发射极电流即为电子的扩散电流。

2. 基区电子与空穴的复合运动形成基极电流

由于基区厚度薄而且掺杂浓度低，即使作为多数载流子，空穴数量也比由发射区扩散过来的电子数量低几个数量级。故基区的空穴只能复合一小部分从发射区扩散到基区的电子，形成基极电流。

3. 基区电子的漂移运动形成集电极电流

由发射区扩散至基区却没被复合的多数电子，与基本身的少数载流子电子一起，在集电结反偏电压的作用下，从基区漂移至集电区，同时集电区的少数载流子空穴也在反偏电压的作用下漂移至基区，形成集电极电流。集电极电流即为反向偏置下 PN 结的漂移电流。与单个反向偏置 PN 结的漂移电流相比，不同的是发射区扩散至基区的电子大大地补充了基区的少数载流子数量，故集电结虽然被反向偏置，漂移电流却很大。另外，在一般情况下，集电区和基区自身提供的少数载流子数量很少，故在近似分析中，可以直接忽略，集电极电流即为基区电子的漂移电流。

工作于放大区的三极管的伏安特性曲线如图 8.4 中注明的"放大区"所示。当基极电流 I_B 固定时，即使集电极-发射极电压 V_{CE} 继续增大，集电极电流 I_C 也不会明显地增大，集电区表现出很大的内阻。

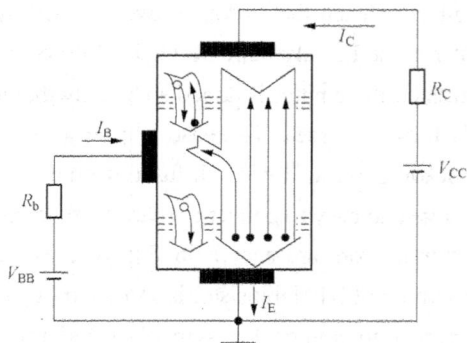

图 8.3 三极管工作在放大区

Fig. 8.3 An NPN BJT operates in the active region

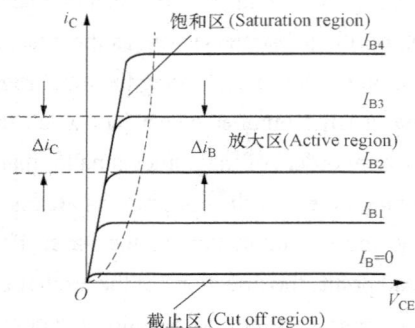

图 8.4 工作于各工作区的三极管伏安特性曲线

Fig. 8.4 The I-V characteristics of a BJT working in the three operation regions

通过上述分析可知，集电极电流为反向偏置的集电结漂移电流，少数载流子数量取决于由发射极扩散过来的电子数量，相比之下集电极自身提供的少数载流子因数量很少可以忽略。而当 V_{CE} 固定时，基极电流 I_B 的变化就会引起集电极电流 I_C 的变化。可见工作于放大区的三极管，集电极电流 I_C 主要由基极电流 I_B 控制，具有放大作用。此时集电极电流 I_C 和基极—发射极电压 V_{BE} 的关系为

$$I_C = I_S e^{V_{BE}/V_T} \tag{8.1}$$

式中，I_S 为饱和电流；V_T 为热电压。

以上分析可见，发射极电流、基极电流和集电极电流还存在一定的关系。假设基极电流 I_B 和集电极电流 I_C 的方向为流入三极管，发射极电流 I_E 流出三极管，如图 8.3 所示。在近似情况下，由于流入和流出三极管的电流相等，有

$$I_E = I_B + I_C \tag{8.2}$$

the current due to the diffusion of the holes can be omitted. Approximately, the emitter current is regarded as the diffusion current of electrons.

2. Base current: the recombination of the electrons and holes

The base is thin and doped so lightly that the major carriers, namely the holes, are several orders of magnitude smaller than the electrons diffused from the emitter. Thus only a fraction of the electrons from the emitter are recombined by the holes in the base, forming the base current.

3. Collector current: the drift of the electrons in the base

With the collector-base junction reverse-biased, the electrons in the base including those from the emitter without recombination and the intrinsic minor carriers of the base itself drift to the collector. At the same time the minor carriers, the holes, from the collector drift to the base as well, both leading to the collector current. That is, the collector current is drift current of a reverse-biased PN junction. But comparing to a single reverse-biased PN junction, because that the electrons diffused from the emitter greatly increase the number of the minor carriers of the base, the collector current is comparably large even though the base-collector junction is reverse-biased. In addition, generally the number of intrinsic minor carriers of the base itself is so small that it can be ignored, thus the collector current is the drift current of the electrons in the base.

The *I-V* characteristic of the transistor operating in the active region is shown in Fig. 8.4, marked with active region. When the base current I_B is fixed, the collector current I_C does not increase significantly though the collector-emitter voltage V_{CE} increases, the collector appearing large resistance. This is because, according to the above analysis, the collector current is essentially the reverse-biased junction current and the number of the minor carriers contributing to the current is determined by the electrons diffusing from the emitter instead of the intrinsic ones provided by the collector itself, which is too small to be noticed. When V_{CE} is fixed, the change in the base current I_B causes the change in the collector current I_C. As we can see in active mode, the collector current I_C is mainly controlled by the base current I_B. That is, the transistor amplifies the current variation. Now the relationship between the collector current I_C and the base-emitter voltage V_{BE} is defined by

$$I_C = I_S e^{V_{BE}/V_T} \tag{8.1}$$

where I_S is the saturation current and V_T is the thermal voltage.

It is also obvious that there are some certain relationships among the emitter current, the base current and the collector current. We assume that the base current I_B and the collector current I_C flow in the transistor while the emitter current I_E flows out as shown in Fig. 8.3. Because the current flowing in the transistor must flows out, approximately, we have

$$I_E = I_B + I_C \tag{8.2}$$

集电极电流 I_C 与基极电流 I_B 的比值为

$$\beta = \frac{I_C}{I_B} \tag{8.3}$$

其中 β 称为共射电流增益。不同用途的三极管 β 值会有所不同，一般情况下，β 的取值为 50～200，有的可高达 1000，有的却只有三四十。影响 β 大小的因素主要有两个：一个是基区厚度，另一个则是发射区和基区的掺杂浓度。

将式（8.2）代入式（8.1）可得发射极电流 I_E 和集电极电流 I_C 之间的关系

$$I_E = \left(\frac{1}{\beta} + 1\right) I_C \tag{8.4}$$

即

$$I_C = \frac{\beta}{1+\beta} I_E = \alpha I_E \tag{8.5}$$

其中

$$\alpha = \frac{\beta}{1+\beta} \tag{8.6}$$

称为共基电流增益。由于 β 通常比较大，故 α 小于 1 但非常接近 1。例如当 $\beta = 100$ 时，$\alpha = 0.99$。

8.1.3　工作在饱和区

当三极管发射结正向偏置，而集电结也正向偏置或虽反向偏置但不至于使基区的少数载流子都漂移至集电区时，三极管处于饱和区，如图 8.4 中注明的"饱和区"所示。此时集电极电流不仅与基极电流 I_B 有关，也与集电极-发射极电压 V_{CE} 有关。

8.1.4　工作在截止区

当三极管发射结和集电结均处于反向偏置时，只有微小的电流流过三极管，此时三极管处于截止状态，其伏安特性曲线如图 8.4 中注明的"截止区"所示。工作在截止区的三极管在数字电路中可以表示低电平。

8.1.5　工作在反向区

当三极管发射结反向偏置，集电结正向偏置，即与工作在放大区情况相反时，三极管处于反向状态，此时可将三极管的集电极和发射极互易使用。但由于集电区与发射区的形状及掺杂浓度上有很大区别，由集电区扩散至基区的多子数量很少，使得共射电流增益 β 大幅降低。仅在一些数字电路中会用到三极管的反向状态。

8.1.6　三极管小结

三极管在构造上为两个背靠背的 PN 结。但与简单 PN 结不同的是，由于三极管基区非常薄，这两个 PN 结处于不同偏置状态时将表现出特殊的特性。三极管的工作状态与发射结及集电结偏置状态关系总结于表 8.1 中。其中当发射结正向偏置而集电结反向偏置时，以 NPN 型管为例，发射区电子在发射结正向偏置的作用下向基区扩散形成发射极电流，这些电子到达基区后除了一小部分与基区的空穴复合形成基极电流外，绝大部分电子在集电结的反向偏置作用下漂移至集电区，形成集电极电流。PNP 型管的工作情况与 NPN 型管类似。此时集电极电流 I_C 受集电极-发射极电压 V_{CE} 的影响不大，却主要受到基极电流 I_B 控制，有一定的倍数关系，具有放大作用。故在模拟电路设计时多将三极管偏置于放大区。

The ratio of the collector current I_C and the base current I_B is defined by

$$\beta = \frac{I_C}{I_B} \tag{8.3}$$

where β is called the common-emitter current gain. Transistors for different purposes have different β values. Generally, β is in the range of 50 to 200, but it can be as high as 1000 or as low as 30 to 40 for special transistors. One of the two main factors affecting β is the base width, and the other is the dope densities of the emitter and the base.

If we substitute (8.2) into (8.1), we have the relationship between the emitter current I_E and the collector current I_C, defined by

$$I_E = \left(\frac{1}{\beta} + 1 \right) I_C \tag{8.4}$$

that is

$$I_C = \frac{\beta}{1+\beta} I_E = \alpha I_E \tag{8.5}$$

where

$$\alpha = \frac{\beta}{1+\beta} \tag{8.6}$$

is called the common-base current gain. Generally, β is large, thus α is less than but very close to unity. For instance, if $\beta = 100$, then $\alpha = 0.99$.

8.1.3   Operation in saturation region

When the emitter-base junction is forward-biased, and the collector-base junction is also forward-biased or reverse-biased so lightly that the minor carriers in the base does not drift to the collector entirely, the transistor then operates in the saturation region as shown in Fig 8.4, marked with saturation region. Now the collector current depends on both the base current I_B and the collector-emitter voltage V_{CE}.

8.1.4   Operation in cut-off region

When both the emitter-base junction and the collector-base junction are reverse-biased, only a tiny current flows through the transistor, and the transistor then operates in the cut-off region as shown in Fig 8.4, marked with cut-off region accordingly. Transistors working in the cut-off region can represent the low level in digital circuits.

8.1.5   Operation in reverse region

When the emitter-base junction is reverse-biased, and the collector-base junction is forward-biased, the transistor then operates in the reverse region. In contrast to the active region, now the collector and the emitter are used conversely. As the emitter and collector are very different both in shape and in dopant density, the common-emitter current gain β is lowered greatly due to less major carriers diffusing from the collector to the base. Only in some specific digital circuits, transistors work in the reverse region.

8.1.6   Summary of transistors

Transistors are formed by two back-to-back PN junctions. Unlike a single PN junction, the two PN junctions do have special features in various bias conditions due to the very thin base region. Different operation modes and their related bias conditions of transistors are summarized in Table 8.1. When the emitter-base junction is forward-biased while the collector-base junction is reverse

表 8.1   三极管不同工作区不同的偏置条件

Table8.1   Bias conditions for different operation regions of BJT

工作状态 （Operation station）	发射结 （BE junction）	集电结 （CE junction）	工作状态 （Operation station）	发射结 （BE junction）	集电结 （CE junction）
放大（Amplified）	正向偏置 （Forward-biased）	反向偏置 （Reverse-biased）	截止（Cut-off）	反向偏置 （Reverse-biased）	反向偏置 （Reverse-biased）
饱和（Saturated）	正向偏置 （Forward-biased）	正向偏置 （Forward-biased）	反向（Reverse）	反向偏置 （Reverse-biased）	正向偏置 （Forward-biased）

8.2 三极管的大信号模型

当三极管工作在放大区时，可以用大信号模型来分析它的直流特性。大信号模型可用于三极管的偏置计算。工作于放大区的 NPN 型三极管大信号模型如图 8.5（a）所示。其中，基极和发射极之间的关系用一个二极管表示，表明当发射结正向偏置时导通，且基极电流 I_B 为

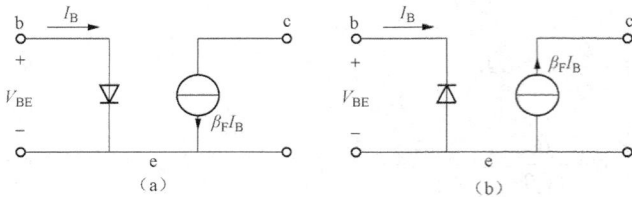

图 8.5 三极管大信号模型

（a）NPN 型三极管；（b）PNP 型三极管

Fig. 8.5   Large-signal model of transistors

(a) NPN transistors; (b) PNP transistors

$$I_B = \frac{I_C}{\beta} = \frac{I_S}{\beta} e^{V_{BE}/V_T} \quad\quad (8.7)$$

集电极和发射极的关系用一个受控电流源表示，表明集电极电流受集电极—发射极电压 V_{CE} 影响不大，即集电区内阻很大，同时集电极电流 I_C 为

$$I_C = \beta I_B \quad\quad\quad\quad\quad\quad\quad\quad (8.8)$$

同理，PNP 型三极管的大信号模型如图 8.5（b）所示。

8.3 三极管放大电路的大信号直流分析

由前面章节可知，当三极管三端加上不同电压时，三极管将工作于不同的区域。一般情况下我们都希望三极管工作于放大区，故需要对三极管进行大信号直流分析。对三极管进行直流分析，即观察当三极管的输入电压从 0 增加至 V_{CC} 时输出电压的变化情况，得出输入—输出电压转移特性曲线。从电压转移特性曲线，可以看出三极管在不同输入情况下所处的工作区域，以及工作区域在输入变化时是如何转换的，从而确定三极管的偏置条件，使三极管处于放大区。

假设 NPN 型三极管接法如图 8.6 所示。这种接法下三极管的发射极同时出现于输入和输出回路中，称为共射接法。我们分析当 V_{BE} 由 0 增加至 V_{CC} 时输出集电极电压 V_{CE} 的变化情况。当 V_{BE} 在 0～0.7V 时，发射结尚未导通，三极管处于截止状态，集电极电流 I_C 接近于 0，此时 R_C 上也基本没有电压降，输出电压 V_{CE} 约为 V_{CC}；当 V_{BE} 超过 0.7V 后，发射结逐渐导通，基极电流 I_B 和集电极电流 I_C 不断增大，此时 R_C 产生电压降，输出电压为 $V_{CE}=V_{CC}-R_C I_C$，并且随着 I_C 增大 V_{CE} 不断下降，曲线陡峭，三极管处于放大区；

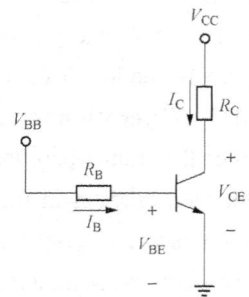

图 8.6 NPN 型三极管的偏置电路

Fig. 8.6   The bias circuit of an NPN transistor

biased, taking NPN for example, the electrons of emitter diffuse to the base, leading to the emitter current. Except that a fraction of the electrons recombine the holes in the base leading to the base current, the majority of the electrons drift to collector with the collector-base junction reverse-biased, resulting in the collector current. PNP counterparts work similarly. Then the collector current I_C is a few times of the base current I_B but not controlled by the collector-emitter voltage V_{CE}, and the transistor amplifies the current. We usually put transistors in the active region in analog designs.

8.2 Large–signal Models of Transistors

The large-signal model of a transistor in active mode is shown in Fig. 8.5 (a). The large-signal behavior of the transistor in the active mode can be analyzed with the help of the large-signal model, which is suitable to determine the bias condition. A diode is modeled between the base and the emitter, denoting that the emitter-base junction is on when forward-biased, and the base current I_B is defined by

$$I_B = \frac{I_C}{\beta} = \frac{I_S}{\beta} e^{V_{BE}/V_T} \tag{8.7}$$

A controlled current source is modeled between the collector and the emitter denoting that the collector-emitter voltage V_{CE} ideally has no influence on the collector current and the collector acts as a high-impedance resistor. The collector current I_C is defined by

$$I_C = \beta I_B \tag{8.8}$$

Similarly, the large-signal model of PNP transistor is shown in Fig. 8.5(b).

8.3 Large–signal DC Analysis of Transistor Amplifiers

As stated in the previous sections, transistors with different bias conditions operate in different regions. Usually we want to put the transistor in the active region, thus the large-signal dc analysis of the transistor is required. The large-signal analysis is to see how the output changes when the input signal varies from 0 to V_{CC} and to plot the transfer characteristic curve. From the transfer characteristic graph, we can see the region in which the transistor operates and the variation of operation region with various input conditions. Then we can figure out the proper bias conditions to put the transistor in active region.

Consider the NPN transistor configuration shown in Fig. 8.6, which is called common-emitter configuration, because the emitter serves as a common terminal between the input and output ports. We are going to see how the collector voltage V_{CE} changes when V_{BE} rises from 0 to V_{CC}. When V_{BE} is below 0.7V, the emitter-base junction is off and the transistor operates in the cut-off region. The collector current I_C is approximately equal to zero. Thus, there is no voltage drop on R_C, and output voltage V_{CE} is roughly equal to V_{CC}. When V_{BE} rises to beyond 0.7V, the emitter-base junction gradually turns on, and the base current I_B and the collector current I_C increase gradually. Now a voltage of $R_C I_C$ drops across R_C and the output voltage is $V_{CE} = V_{CC} - R_C I_C$. The transistor enters the active region and as I_C increases, V_{CE} decreases steeply. When V_{BE} is high enough, the voltage drop across R_C will

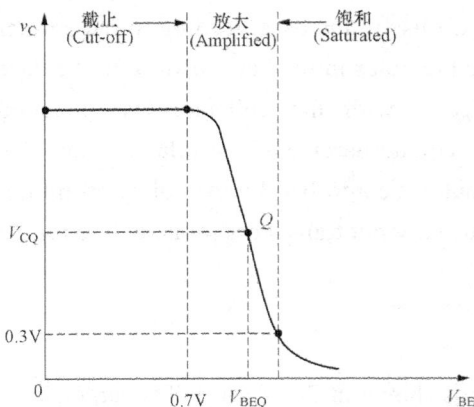

图 8.7　输入输出电压转移特性曲线

Fig. 8.7　The transfer characteristic curve

当 V_{BE} 增大至一定电压后，I_C 在 R_C 上产生的电压降增大，使得集电极电压 V_{CE} 低于 0.3V，此时由于集电结反向电压过小，三极管进入饱和区。连续的输入—输出电压转移特性曲线如图 8.7 所示。由图可见，随着输入电压由 0 向 V_{CC} 增加，三极管先从截止区进入放大区再到饱和区。同时可见，在共射接法下，当输入电压升高时，输出电压降低，即输入和输出相位相反。

当 V_{BE} 确定时，V_{CE} 也随之确定于某个电压，这个点称为静态工作点，或简称 Q 点。正常情况下应使 Q 点落入放大区中。

利用大信号模型可定量得出三极管的静态工作点。图 8.6 所示电路的大信号模型等效电路图如图 8.8 所示。根据二极管的相关知识和基尔霍夫电压定律，在输入回路和输出回路有

$$I_B = \frac{V_{BB} - 0.7}{R_B} \tag{8.9}$$

$$I_C = \beta I_B \tag{8.10}$$

$$V_E = V_{CC} - I_C R_C \tag{8.11}$$

一般情况下令 $V_{CE} = \frac{1}{2} V_{CC}$，使三极管工作在放大区。可通过上述各式计算出所需的电阻大小。除此之外，还可以利用图解法来得出三极管的静态工作点。因为有

$$V_{BE} = V_{BB} - I_B R_B \tag{8.12}$$

$$V_{CE} = V_{CC} - I_C R_C \tag{8.13}$$

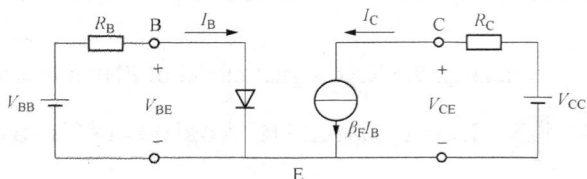

图 8.8　三极管大信号模型等效电路图

Fig. 8.8　The equivalent circuit of a transistor's large signal model

故可以分别在三极管的输入和输出特性曲线中分别作斜率为 $-1/R_B$ 和 $-1/R_C$ 的直线，这两条直线分别交横轴于点 $(V_{BB}, 0)$ 和 $(V_{CC}, 0)$，如图 8.9 所示。直线与输入或输出特性曲线的交点为以上方程的解，也是静态工作点。

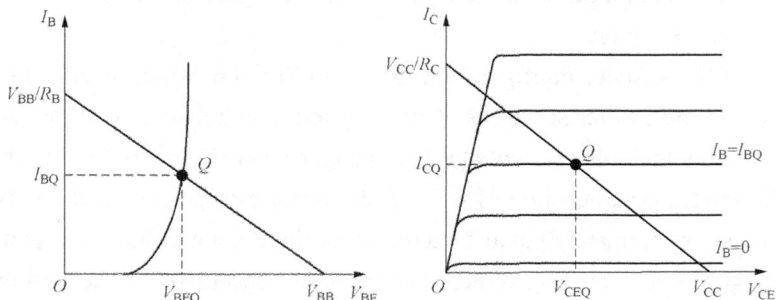

图 8.9　分别通过输入和输出特性曲线确定静态工作点

Fig. 8.9　Determine the Q point using the input and output I-V characteristic curves respectively

be so large that the collector voltage V_{CE} decreases to below 0.3 V. In this case, the transistor enters the saturation region due to the low collector-base junction reverse voltage. The transfer characteristic curve is shown in Fig. 8.7 It is obvious that as the input rises from 0 to V_{CC}, the transistor enters the cut-off region first, then enters the active region, and finally enters the saturation region. It is also obvious that under the common-emitter configuration, the input voltage increases whereas the output voltage decreases; in other words, there is a 180° phase difference between the input and output signals. The point at which V_C is determined by a certain V_{BE} is called the quiescent point, or Q point for short. Normally the Q point is set in the active region.

The quiescent point can be figured out quantitatively by the large-signal model. The equivalent circuit of Fig. 8.6 is shown in Fig. 8.8. According to the knowledge of diode and KVL, in the input and the output loop we can obtain

$$I_B = \frac{V_{BB} - 0.7}{R_B} \tag{8.9}$$

$$I_C = \beta I_B \tag{8.10}$$

$$V_{CE} = V_{CC} - I_C R_C \tag{8.11}$$

Usually, we set $V_{CE} = \frac{1}{2} V_{CC}$ to bias the transistor in the active region. By the equations above the proper resistors can be determined. Besides, the quiescent point can also be figured out graphically. Since

$$V_{BE} = V_{BB} - I_B R_B \tag{8.12}$$

$$V_{CE} = V_{CC} - I_C R_C \tag{8.13}$$

we can draw lines with the slope of $-1/R_B$ and $-1/R_C$ on the I-V characteristic curves of the input and the output, crossing the horizontal axes at the points $(V_{BB}, 0)$ and $(V_{CC}, 0)$, respectively, as shown in Fig. 8.9. The crossing points are the solutions of the equations above and are also the quiescent points.

William Shockley (13 February, 1910 - 12 August, 1989) was an American physicist and inventor. He was the manager of a research group that included John Bardeen and Walter Brattain. The three scientists invented the point-contact transistor in 1947 and were jointly awarded the 1956 Nobel Prize in Physics. Shockley also invented a different sort of transistor based on junctions instead of point contacts. His attempts to commercialize the junction transistor design in the 1950s and 1960s led to California's "Silicon Valley" becoming a hotbed of electronics innovation.

8.4 三极管放大电路的小信号等效电路

8.4.1 三极管的小信号放大原理

当三极管被偏置至放大区后，就可以按图 8.10（a）所示在共射极接法的三极管输入端叠加一个微小信号 v_b 进行放大。微小信号能得到放大的原理是，从三极管的输入—输出电压转移特性曲线可见，在放大区中，v_{BE} 微小的变化能引起 v_{CE} 大幅度的相应变化，如图 8.10（b）所示。

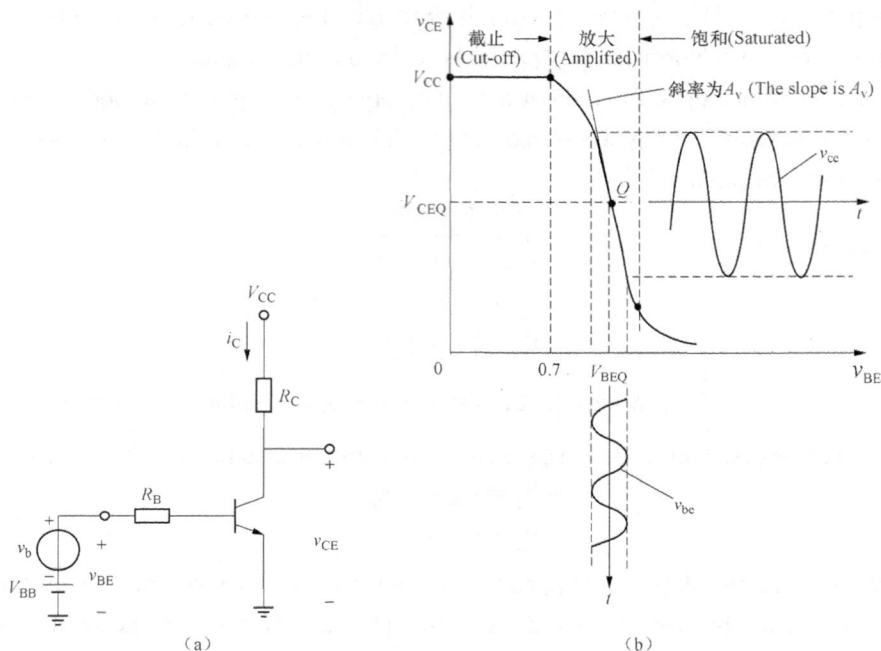

图 8.10 三极管的小信号放大电路及小信号放大原理
（a）小信号放大电路；（b）小信号放大原理
Fig. 8.10 A small-signal amplifier circuit of the transisfor and its principle
(a) A small-signal amplifier circuit; (b) Principle of small-signal amplification

如果 Q 点设置得过高，即静态工作点 V_{BEQ} 过低，当输入信号为负半周时，三极管可能进入截止区，此时输出出现截止失真，表现为输出电压 v_{ce} 顶部被削平，如图 8.11（a）所示。如果 Q 点设置得过低，即静态工作点 V_{BEQ} 过高，当输入信号为正半周时，三极管可能进入饱和区，此时输出出现饱和失真，表现为输出电压 v_{ce} 底部被削平，如图 8.11（b）所示。故在设计时应注意合理选择静态工作点，防止三极管离开放大区而出现失真。

但大多时候不希望直接将微小信号源与直流偏置电压串联，只对输出的交流变化部分感兴趣。此时可以利用电容将微小输入信号耦合叠加至三极管的输入，也可利用电容将输出的直流部分去掉，获得交流部分，如图 8.12 所示。

8.4.2 直流通路与交流通路

由上一节的分析得知，加在三极管上的电压可看成直流偏置电压和交流电压的线性叠加。于是可以将整个三极管放大电路分成两个电路的线性叠加：一个为直流通路，用于分析放大电路的直流偏置状态；另一个为交流通路，用于分析交流信号的放大情况。由于电容及电感元件在直流和交流条件下有不同的反应，将电路分为直流通路和交流通路的一般方法如下：

8.4 Small–Signal Equivalent Circuits of Transistors

8.4.1 Principle of small-signal amplification

When the transistor is biased in the active region with the common-emitter configuration, as shown in Fig. 8.10 (a), we can superpose at the input node a small signal v_b to amplify. The principle of small signal amplification is that, in the active region, even slight fluctuations in v_{BE} can change v_{CE} significantly, as we can see from the transfer characteristic graph shown in Fig. 8.10 (b).

Suppose the Q point is set too high, i.e., the value of V_{BEQ} is too small. When the input signal goes negative, the transistor may enter the cut-off region and the top of the output waveform of v_{ce} is trimmed, as shown in Fig. 8.11 (a). On the other hand, suppose the Q point is set too low, i.e., the value of V_{BEQ} is too large. When the input signal goes positive, the transistor may enter the saturation region and the bottom of the output waveform of v_{ce} is cut, as shown in Fig. 8.11 (b). In both cases distortion occurs. Thus, the quiescent point should be set carefully to avoid distortion.

Usually we do not want to connect the signal source to the DC biasing source directly in series and we are merely interested in the AC component of the output. So capacitors can be used to couple the signal at the input terminal of the transistor and to filter out the DC component at the output as shown in Fig. 8.12.

8.4.2 Large-signal DC circuit and small-signal AC circuit

As stated in the previous section, the output of the transistor is the summation of the responses due to the DC bias source and the ac signal source. Therefore, the original circuit can be seen as the superposition of such two separate circuits: one is the DC bias circuit used to analyze the Q point, and the other is the small-signal ac circuit, used to analyze the response of the ac input signal. Due to the different responses of capacitors and inductors to DC and ac signals, here are the guidelines of how to separate the original circuit into large-signal DC bias circuit and small-signal ac circuits:

More facts about biasing

Biasing in electronics means establishing predetermined voltages or currents at various points of an electronic circuit for the purpose of establishing proper operating conditions in electronic components. Many electronic devices such as transistors and vacuum tubes, whose function is processing AC signals, also require a DC current or voltage to operate correctly. To achieve this, a bias circuit is built. The AC signal applied to the device is superposed on this DC bias current or voltage. The operating point of a device, also known as quiescent point, or Q-point, is the DC voltage or current at a specified terminal of the device with no input signal applied.

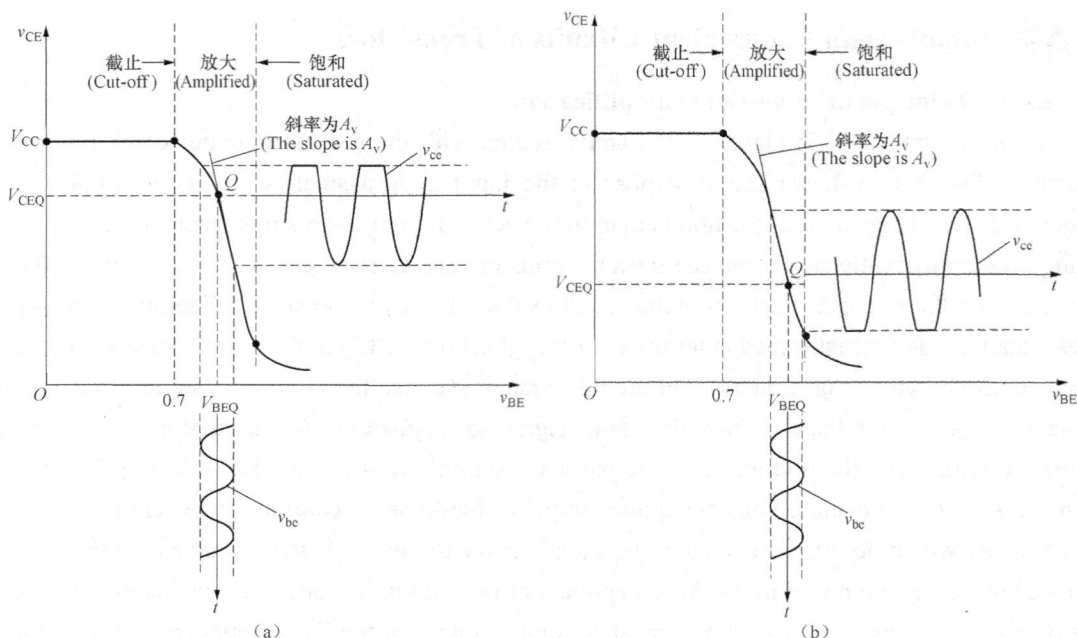

图 8.11　放大电路的失真

（a）三极管进入截止区产生失真；（b）三极管进入饱和区产生失真

Fig. 8.11　The distortion of the amplifier circuit

(a) The transistor enters the cut-off region and causes distortion; (b) The transistor enters the saturation region and causes distortion

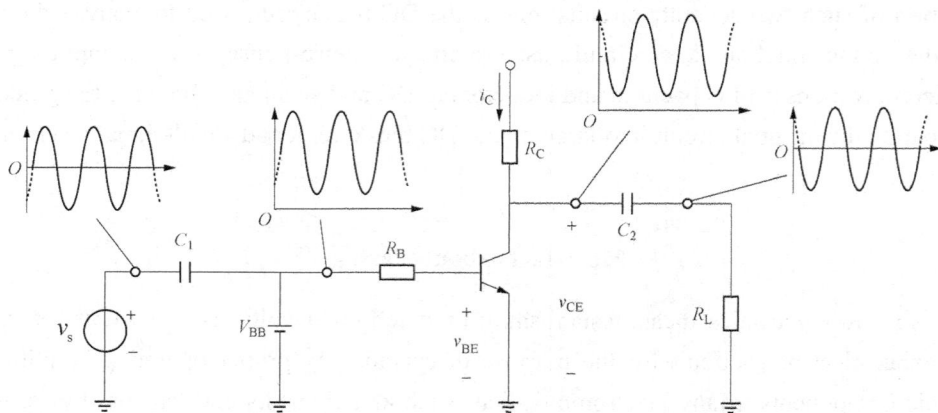

图 8.12　通过电容将微小输入信号耦合叠加至三极管的输入

Fig. 8.12　Use capacitors to couple the signal at the input terminal of the transistor

（1）在直流通路中，由于不考虑交流信号，因此可以将电容做开路处理，电感做短路处理，信号电流源也开路，而信号电压源短路，但电源都保留内阻；

（2）在交流通路中，由于不考虑直流信号，因此可以将电容短路，电感开路，直流电流源开路、直流电压源短路，但保留内阻。

根据上述方法，图 8.12 所示的放大电路的直流通路和交流通路分别如图 8.13（a）和图 8.13（b）所示，其中 r_{sig} 表示信号源 v_{sig} 的内阻。这里忽略了直流电压源的内阻。

(1) The large-signal DC bias equivalent circuit can be constructed by replacing capacitors with open circuits, inductors with short circuits, AC current sources with open circuits and, AC voltage sources with short circuits and their internal resistances;

(2) The small-signal AC equivalent circuit or small-signal equivalent circuit can be constructed by replacing capacitors with short circuits, inductors with open circuits, DC current sources with open circuits and, DC voltage sources with short circuits and their internal resistances.

By the guidelines mentioned above, the large-signal DC bias circuit and the small-signal AC circuit of the amplifier in Fig. 8.12 are shown in Fig. 8.13(a) and Fig. 8.13(b), respectively, where the internal resistance of the AC source v_{sig} is denoted by r_{sig} and the internal resistance of the dc source is ignored.

John Bardeen (23 May, 1908 – 30 January, 1991) was an American physicist and electrical engineer, the only person to have won the Nobel Prize in Physics twice: first in 1956 with William Shockley and Walter Brattain for the invention of the transistor; and again in 1972 with Leon Cooper and John Robert Schrieffer for a fundamental theory of conventional superconductivity known as the BCS theory.

Working in Bell Laboratories, Bardeen and his colleagues succeeded in creating a point-contact transistor that achieved amplification. The transistor revolutionized the electronics industry, allowing the Information Age to occur, and made possible the development of almost every modern electronic device, from telephones to computers to missiles.

Bardeen's developments in superconductivity, which won him his second Nobel, are used in Nuclear Magnetic Resonance Spectroscopy (NMR) or its medical sub-tool magnetic resonance imaging (MRI).

8.4.3 三极管的小信号等效电路

如果忽略三极管放大电路的直流偏置情况，只关注它在交流情况下输入与输出的关系，并假设输入信号足够小使得输出信号与它呈线性关系变化，可将图 8.13（b）中的点画线框抽象成一个黑箱子，且这个箱子里面的电路是线性的。用小信号模型替换三极管则可达到这个目的。低频情况下的三极管小信号模型如图 8.14 所示，它将三极管在感兴趣的静态工作点做线性化处理，只包含线性元件，大大简化了三极管放大电路的分析。

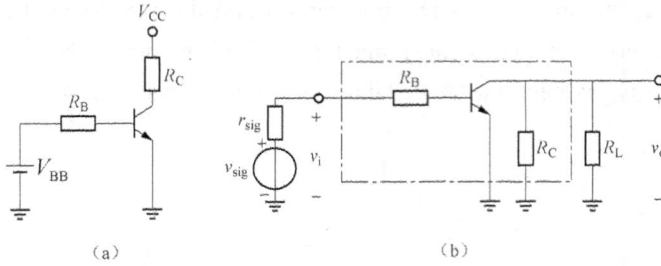

图 8.13 图 8.12 所示放大电路的直流通路与交流通路
（a）直流通路；（b）交流通路

Fig. 8.13 The large-signal DC bias circuit and the small-signal AC circuit of the amplifier circuit shown in Fig. 8.12

(a) The large-signal DC bias circuit; (b) The small-signal AC circuit

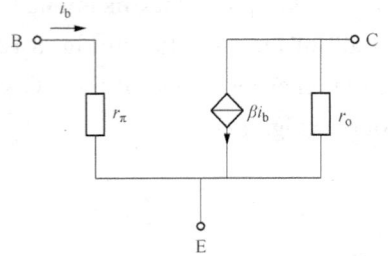

图 8.14 低频情况下的三极管小信号模型

Fig. 8.14 The small-signal model of transistors at low frequencies

低频情况下的三极管小信号模型包含三个元件：一个是输入回路的输入电阻 r_π，另外两个分别是输出回路的受控电流源及输出电阻 r_o。以下将逐一分析它们的含义及取值。

（1）输入电阻 r_π。从图 8.15（a）的输入特性曲线可见，在静态工作点附近，若将基极电流 I_B 随基极-发射极电压 V_{BE} 的变化近似呈线性关系，则可在输入特性曲线 Q 点作一条切线，这条切线的斜率就是输入回路电阻 r_π 的倒数 $1/r_\pi$。根据式（8.7）有

$$\frac{1}{r_\pi} = \frac{\partial I_B}{\partial V_{BE}} = \frac{\partial \left(\dfrac{I_S}{\beta} e^{V_{BE}/V_T} \right)}{\partial V_{BE}} = \frac{I_S}{\beta} e^{V_{BE}/V_T} \cdot \frac{1}{V_T} = \frac{I_C}{V_T} \tag{8.14}$$

即

$$r_\pi = \frac{V_T}{I_C} \tag{8.15}$$

式中，V_T 为热电压，常温下约为 26mV。

可见输入电阻与静态偏置时的集电极电流 I_C 有关，I_C 越大则 r_π 越小。

（2）受控电流源。由输出特性曲线可知，集电极电流 i_c 主要受到基极电流 i_b 控制，且有 $i_c = \beta i_b$，故可用一个输出为 βi_b 的受控电流源表示集电极电流。

（3）输出电阻 r_o。从图 8.15（b）所示的输出特性曲线可见，虽然进入放大区后集电极电流 i_C 随集电极-发射极电压 V_{CE} 变化平坦，但仍然有一定斜率。同理在输出特性曲线 Q 点作一条切线，这条切线的斜率就是输出回路电阻 r_o 的倒数 $1/r_o$。输出电阻 r_o 源于基区宽度调制效应，在此不详细介绍。输出电阻 r_o 一般非常大，取值可达数十千欧姆。将图 8.13（b）中的三极管替换成小信号等效电路后，得到如图 8.16 所示的电路。

8.4.3   Small-signal equivalent circuits of transistors

If we are interested merely in the ac input and output, ignore the dc bias component, and assume that the input signal is weak enough that the output nonlinear distortion is negligible, we can replace the dashed box in Fig. 8.13(b) with a linear circuit. The linear circuit is implemented with the small-signal model of transistors. The small-signal model of transistors at low frequencies is shown in Fig. 8.14, in which only linear elements are included. The behaviors of the transistor at the quiescent point are linearized, eliminating the difficulties of analysis considerably.

The small-signal model of transistors at low frequencies includes three elements, the input resistance r_π in the input loop, the controlled current source, and output resistance r_o in the output loop. We are going into the details of these parameters.

(1) Input resistance r_π. As shown in the input transfer characteristics in Fig. 8.15 (a) (The input transfer characteristics), around the quiescent point, if the variation of the base current I_B due to the change of the emitter-base voltage V_{BE}, is approximate to be linear, a tangent line with a slope of $1/r_\pi$, which is the reciprocal of the input resistance r_π, can be drawn across the Q point. From (8.7) we have

$$\frac{1}{r_\pi} = \frac{\partial I_B}{\partial V_{BE}} = \frac{\partial \left(\dfrac{I_S}{\beta} e^{V_{BE}/V_T} \right)}{\partial V_{BE}} = \frac{I_S}{\beta} e^{V_{BE}/V_T} \cdot \frac{1}{V_T} = \frac{I_C}{V_T} \tag{8.14}$$

that is

$$r_\pi = \beta \frac{V_T}{I_C} \tag{8.15}$$

where V_T is the thermal voltage, and approximates to be 26mV at room temperature.

It shows that the input resistance is related to the quiescent collector current, I_C. The higher the I_C, the lower the input resistance r_π.

(2) Controlled current source. From the transfer characteristics curve, it shows that the collector current i_c is mainly controlled by the base current i_b, and $i_c = \beta i_b$. Thus the collector current is modeled by a controlled current source of βi_b.

(3) Output resistance r_o. Though in the active region the collector current i_C changes along with the variation of the collector-emitter voltage V_{CE} insignificantly, there is still a slope for the i_C curve in the transfer characteristics graph. Similarly, a tangent line across the Q point with a slope of $1/r_o$, which is the reciprocal of the output resistance r_o, can be drawn. The output resistance r_o is resulted from the Early effect, which can generally be as large as tens of kilo ohms.

The small-signal equivalent circuit of Fig. 8.13(b) after replacing the transistor with its small-signal model is shown in Fig. 8.16.

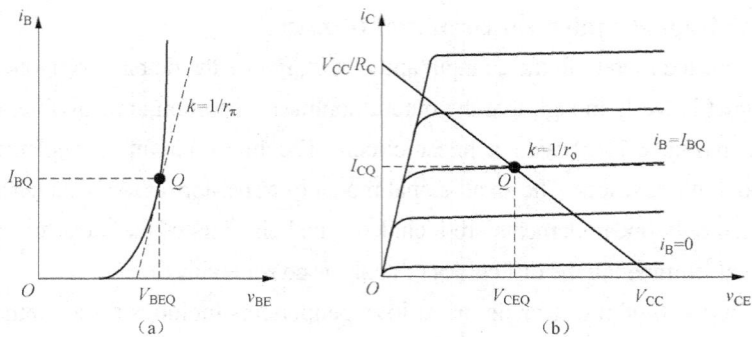

图 8.15 放大电路的特性曲线

（a）输入特性曲线；（b）输出特性曲线

Fig. 8.15 The transfer characteristics of the amplifier

(a) The input transfer characteristics; (b) The output transfer characteristics

讨论： 几种常见的三极管放大电路的实际应用

1. 共射放大器

共射放大器的电路图如图 8.17 所示，请同学们：

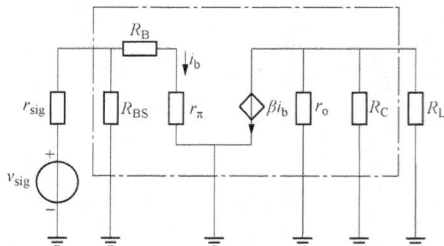

图 8.16 图 8.13（b）的小信号等效电路

Fig. 8.16 The small-signal equivalent circuit of Fig. 8.13(b)

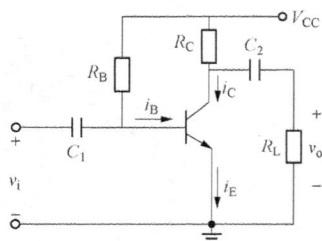

图 8.17 共射放大器

Fig. 8.17 Common-emitter amplifier

（1）从电路图中找出直流通路和交流通路，注意电容的作用。

（2）对电路进行大信号分析，确定静态工作点。

（3）画出小信号等效电路图，计算放大倍数、输入阻抗和输出阻抗。

分析：（1）对于直流情况，电容开路处理，交流信号源短路处理，但是保留内阻，因此，其直流通路为图 8.18（a）。

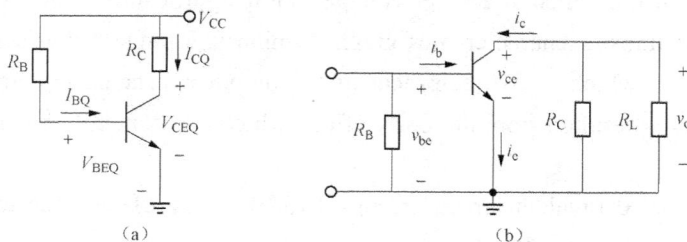

图 8.18 共射放大器的直流通路与交流通路

（a）直流通路；（b）交流通路

Fig. 8.18 The large-signal DC circuit and the small-signal AC circuit of the common-emitter amplifier

(a) The large-signal DC circuit; (b) The small-signal AC circuit

Discussion: Practical applications of transistor amplifier circuits

1. Common-emitter amplifier

The circuit of a common-emitter amplifier is shown in Fig. 8.17.

(1) Please find out both the dc bias circuit and ac circuit, and analyze the function of the capacitors;

(2) Please analyze the dc bias circuit and determine the Q point;

(3) Please draw the small-signal equivalent circuit and calculate its gain, input impedance and output impedance.

Analysis:

(1) When constructing the dc bias circuit, the capacitors are replaced by open circuits and the AC voltage source is replaced by short circuits (its internal resistance is kept). Thus the dc bias circuit is shown in Fig. 8.18(a).

Walter Brattain (10 February, 1902 – 13 October, 1987) was an American physicist at Bell Labs who, along with fellow scientists John Bardeen and William Shockley, invented the point-contact transistor in 1947. They shared the 1956 Nobel Prize in Physics for their invention. Bardeen was a quantum physicist, Brattain was a gifted experimenter in materials science, and Shockley, the leader of their team, was an expert in solid-state physics. Brattain devoted much of his life to research on surface states.

（2）对于交流小信号而言，电容相当于短路。直流偏置电源相对于交流信号而言也相当于短路，则电阻 R_B 和 R_C 的上端相当于接地。因此交流通路为图 8.18（b）。

（3）对直流通路进行分析可知，静态工作点 Q 满足

$$\begin{cases} I_{BQ} = \dfrac{V_{BB} - V_{BEQ}}{R_B} \\ I_{CQ} = \beta I_{BQ} \\ V_{CEQ} = V_{CC} - I_{CQ}R_C \end{cases} \tag{8.16}$$

（4）画出小信号等效电路图如图 8.19 所示。

（5）求放大电路的动态参数。

1）电压增益。计算该放大电路的电压增益即确定输出电压与输入电压的比值。分析图8.19 有

$$\begin{cases} i_b = \dfrac{v_i}{R_B + r_\pi} \\ i_c = \beta i_b \\ v_o = -i_c R_C \end{cases} \tag{8.17}$$

所以电压增益为

$$A_v = \frac{v_o}{v_i} = \frac{-i_c R_C}{i_b (R_B + r_\pi)} = -\frac{\beta R_C}{R_B + r_\pi} \tag{8.18}$$

2）输入电阻。

$$r_i = \frac{v_i}{i_b} = R_B + r_\pi \tag{8.19}$$

3）输出电阻。

$$r_o = R_C \tag{8.20}$$

2. 共基放大电路

（1）从图 8.20 中找出直流通路和交流通路。

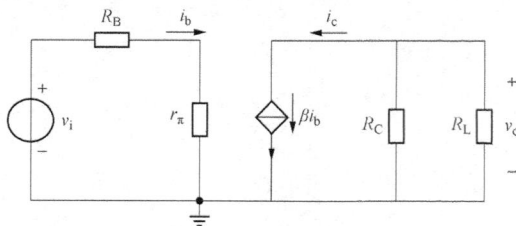

图 8.19　小信号等效电路
Fig. 8.19　Small-signal equivalent circuit

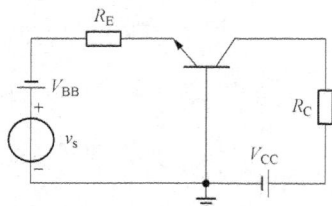

图 8.20　共基放大电路
Fig. 8.20　Common-base amplifier

（2）对电路进行大信号分析，确定静态工作点。

（3）画出小信号等效电路图，计算放大倍数、输入阻抗和输出阻抗。

(2) For an ac signal, both the capacitors and the dc bias voltage sources response as short circuits. Therefore, the upper terminals of resistances R_B and R_C are both grounded. The small-signal ac circuit is shown in Fig. 8.18(b).

(3) According to the dc bias circuit, the Q point is defined by

$$\begin{cases} I_{BQ} = \dfrac{V_{BB} - V_{BEQ}}{R_B} \\ I_{CQ} = \beta I_{BQ} \\ V_{CEQ} = V_{CC} - I_{CQ} R_C \end{cases} \qquad (8.16)$$

(4) Replacing the transistor with its small-signal model, the small-signal equivalent circuit is shown in Fig. 8.19.

(5) The parameters of voltage gain, input resistance, and output resistance are calculated as follows.

1) Voltage gain. The voltage gain is the ratio of the output voltage and the input voltage. Referring to Fig 8.19, we have

$$\begin{cases} i_b = \dfrac{v_i}{R_B + r_\pi} \\ i_c = \beta i_b \\ v_o = -i_c R_C \end{cases} \qquad (8.17)$$

That is, the voltage gain

$$A_v = \frac{v_o}{v_i} = \frac{-i_c R_C}{i_b (R_B + r_\pi)} = -\frac{\beta R_C}{R_B + r_\pi} \qquad (8.18)$$

2) Input resistance

$$r_i = \frac{v_i}{i_b} = R_B + r_\pi \qquad (8.19)$$

3) Output resistance

$$r_o = R_C \qquad (8.20)$$

2. Common-base amplifier

(1) Find out the dc bias circuit and the small-signal ac circuit from Fig. 8.20;

(2) Perform the large-signal dc analysis and determine the Q point;

(3) Draw the small-signal equivalent circuit and calculate the voltage gain, the input resistance and the output resistance.

分析：（1）同前面分析类似，共基放大电路的直流通路为图 8.21（a），而交流通路为图 8.21（b）所示。

（2）分析直流通路，得其静态工作点 Q 满足

$$
\begin{cases}
I_{EQ} = \dfrac{V_{BB} - V_{BEQ}}{R_E} \\[2mm]
I_{CQ} = \dfrac{\beta}{1+\beta} I_{EQ} \\[4mm]
V_{CEQ} = V_{CC} - I_{CQ} R_C + V_{BEQ}
\end{cases}
\tag{8.21}
$$

（3）画出小信号等效电路如图 8.22 所示。

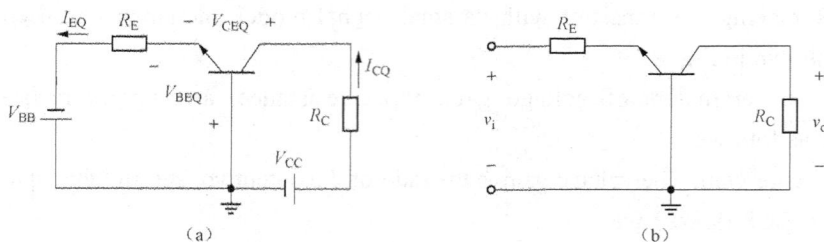

图 8.21　共基放大电路的直流通路与交流通路

（a）直流通路；（b）交流通路

Fig. 8.21　The large-signal DC circuit and the small-signal AC circuit of the common-base amplifier

(a) The large-signal DC circuit; (b) The small-signal AC circuit

（4）求放大电路的动态参数。

1）电压增益为

$$
A_v = \frac{v_o}{v_i} = \frac{i_c R_C}{i_e R_E + i_b r_\pi} = \frac{\beta i_b R_C}{(1+\beta) i_b R_E + i_b r_\pi} = \frac{\beta R_C}{(1+\beta) R_E + r_\pi}
\tag{8.22}
$$

2）输入电阻

$$
r_i = \frac{v_i}{i_e} = \frac{i_e R_E + i_b r_\pi}{i_e} = R_E + \frac{r_\pi}{1+\beta}
\tag{8.23}
$$

3）输出电阻

$$
r_o = R_C
\tag{8.24}
$$

3. 共集放大电路（射极跟随器）

（1）从图 8.23 中找出直流通路和交流通路。

（2）对电路进行大信号分析，确定静态工作点。

（3）画出小信号等效电路图，计算放大倍数、输入阻抗和输出阻抗。

Analysis:

(1) Similarly, for ac signal, both the capacitors and the dc voltage sources response as short circuits. The dc bias circuit and the small-signal ac circuit of the common-base amplifier are shown in Fig. 8.21(a) and Fig. 8.21(b) respectively.

(2) According to the dc bias circuit, the Q point is defined by

$$\begin{cases} I_{EQ} = \dfrac{V_{BB} - V_{BEQ}}{R_E} \\[3mm] I_{CQ} = \dfrac{\beta}{1+\beta} I_{EQ} \\[3mm] V_{CEQ} = V_{CC} - I_{CQ} R_C + V_{BEQ} \end{cases} \tag{8.21}$$

(3) Replacing the transistor with its small-signal model, the small-signal equivalent circuit is shown in Fig. 8.22.

(4) The parameters of voltage gain, input resistance, and output resistance are calculated as follows.

1) Voltage gain

$$A_v = \frac{v_o}{v_i} = \frac{i_c R_C}{i_e R_E + i_b r_\pi} = \frac{\beta i_b R_C}{(1+\beta) i_b R_E + i_b r_\pi} = \frac{\beta R_C}{(1+\beta) R_E + r_\pi} \tag{8.22}$$

2) Input resistance

$$r_i = \frac{v_i}{i_e} = \frac{i_e R_E + i_b r_\pi}{i_e} = R_E + \frac{r_\pi}{1+\beta} \tag{8.23}$$

3) Output resistance

$$r_o = R_C \tag{8.24}$$

Common-collector amplifier (emitter follower):

(1) Find out the dc bias circuit and the small-signal ac circuit from Fig. 8.23.

(2) Perform the large-signal dc analysis and determine the Q point.

(3) Draw the small-signal equivalent circuit and calculate the voltage gain, the input resistance and the output resistance.

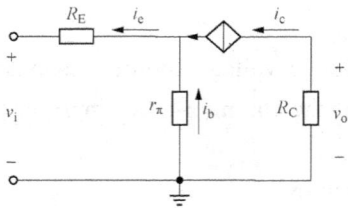

图 8.22 小信号等效电路

Fig. 8.22 Small-signal equivalent circuit

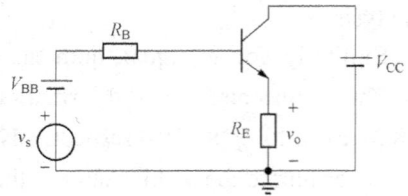

图 8.23 共集放大电路（射极跟随器）

Fig. 8.23 Common-collector amplifier (emitter follower)

分析：（1）同前面分析类似，共集放大电路的直流通路为图 8.24（a），而交流通路为图 8.24（b）所示。

（2）分析直流通路，得其静态工作点为

$$\begin{cases} I_{BQ} = \dfrac{V_{BB} - V_{BEQ}}{R_B + (1+\beta)R_E} \\ I_{EQ} = (1+\beta)I_{BQ} \\ V_{CEQ} = V_{CC} - I_{EQ}R_E \end{cases} \qquad (8.25)$$

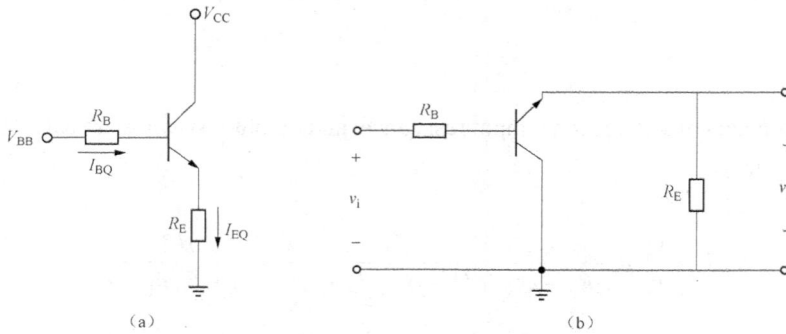

（a） （b）

图 8.24 共集放大电路（射极跟随器）的直流通路和交流通路

（a）大信号直流通路；（b）小信号交流通路

Fig. 8.24 The large-signal DC circuit and the small-signal AC circuit of a common-collector amplifier (emitter follower)

(a) The large-signal DC circuit; (b) The small-signal AC circuit

（3）画出小信号等效电路如图 8.25 所示。

（4）求放大电路的动态参数。

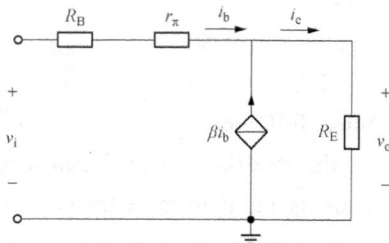

图 8.25 小信号等效电路

Fig. 8.25 Small-signal equivalent circuit

Analysis:

(1) The dc bias circuit and the small-signal ac circuit of the common-collector amplifier are shown in Fig. 8.24(a) and Fig. 8.24(b), respectively.

(2) According to the dc bias circuit, the Q point is defined by

$$\begin{cases} I_{BQ} = \dfrac{V_{BB} - V_{BEQ}}{R_B + (1+\beta)R_E} \\[3mm] I_{EQ} = (1+\beta)I_{BQ} \\[3mm] V_{CEQ} = V_{CC} - I_{EQ}R_E \end{cases} \tag{8.25}$$

(3) Replacing the transistor with its small-signal model, the small-signal equivalent circuit is shown in Fig. 8.25 (Small-signal equivalent circuit).

(4) The parameters of voltage gain, input resistance, and output resistance are calculated as follows.

1) Voltage gain is

$$A_v = \frac{v_o}{v_i} = \frac{i_e R_E}{i_e R_E + i_b(r_\pi + R_B)} = \frac{(1+\beta)i_b R_C}{(1+\beta)i_b R_E + i_b(r_\pi + R_B)} = \frac{(1+\beta)R_C}{(1+\beta)R_E + r_\pi + R_B} \tag{8.26}$$

2) Input resistance

$$r_i = \frac{v_i}{i_e} = \frac{i_e R_E + i_b(R_B + r_\pi)}{i_b} = R_B + r_\pi + (1+\beta)R_E \tag{8.27}$$

More facts about transistors

The most commonly used FET is the MOSFET. The main advantage of a MOSFET over a regular transistor is that it requires very little current to turn on (less than 1 mA), while delivering a much higher current to a load (10 to 50 A or more).

MOSFETs are widely applied in digital integrated circuits, such as microprocessors and memory devices, which contain thousands to millions of integrated MOSFETs on each device. MOSFETs provide the basic switching functions required to implement logic gates and data storage. Discrete devices are widely used in applications such as switch mode power supplies, variable-frequency drives and other power electronics applications where each device may be switching hundreds or thousands of watts.

1）电压增益为

$$A_v = \frac{v_o}{v_i} = \frac{i_e R_E}{i_e R_E + i_b (r_\pi + R_B)} = \frac{(1+\beta)i_b R_C}{(1+\beta)i_b R_E + i_b(r_\pi + R_B)} = \frac{(1+\beta)R_C}{(1+\beta)R_E + r_\pi + R_B} \qquad (8.26)$$

2）输入电阻

$$r_i = \frac{v_i}{i_e} = \frac{i_e R_E + i_b (R_B + r_\pi)}{i_b} = R_B + r_\pi + (1+\beta)R_E \qquad (8.27)$$

3）输出电阻的求解比较复杂，如图 8.26 所示，首先将独立源置零，然后在输出端加上测试电压 v_x，求出测试电流 i_x，则输出电阻可由 $r_o = v_x / i_x$ 得到。下面将求解 v_x / i_x。

$$\begin{cases} i_b = \dfrac{v_x}{R_B + r_\pi} \\[2mm] i_e = (1+\beta)i_b \\[2mm] i_x + i_e = \dfrac{v_x}{R_E} \end{cases} \qquad (8.28)$$

所以

$$r_o = \frac{v_x}{i_x} = \frac{v_x}{\dfrac{v_x}{R_E} - (1+\beta)\dfrac{v_x}{R_B + r_\pi}} = \frac{1}{\dfrac{1}{R_E} + (1+\beta)\dfrac{1}{R_B + r_\pi}} \qquad (8.29)$$

即

$$r_o = R_E // \frac{R_B + r_\pi}{1+\beta} \qquad (8.30)$$

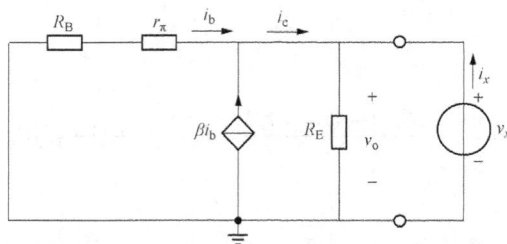

图 8.26　求输出电阻的电路图

Fig. 8.26　Circuit diagram used to calculate the output resistance

小结： 三种形式的放大电路的比较

（1）共射放大电路具有较大的电压放大倍数和电流放大倍数，输入电阻适中，输出电阻较大，但是频带较窄。因而常用在低频放大电路的输入级、中间级和输出级。除此之外，该结构的输出信号与输入信号是反向的。

（2）共基放大电路只能放大电压信号，输入电阻较小，放大倍数与共射放大电路相当，但是频率特性非常好，是三种形式中高频特性最好的，常用于宽频带放大电路。

（3）共集放大电路电压放大倍数略小于 1，具有电压跟随的性质，故又称为电压跟随器，但是有电流放大的作用，因而仍可以进行功率放大。共集放大电路的输入电阻最大（对电流的索取能力小），输出电阻最小（带负载能力强），故常常用于阻抗转化，常用于多级放大电路的输入级和输出级。除此之外，该结构的输出信号与输入信号是同向的了，因此电流放大倍数是个正值。

3) The calculation of the output resistance needs more efforts. As illustrated in Fig. 8.26 (Circuit diagram used to calculate the output resistance), we short the input, inject a test signal v_x at the output node, and obtain the test output current i_x. Then the output resistance is calculated as $r_o = v_x / i_x$. Let us solve for v_x / i_x

$$\begin{cases} i_b = \dfrac{v_x}{R_B + r_\pi} \\ i_e = (1 + \beta)i_b \\ i_x + i_e = \dfrac{v_x}{R_E} \end{cases} \tag{8.28}$$

we can get

$$r_o = \frac{v_x}{i_x} = \frac{v_x}{\dfrac{v_x}{R_E} - (1 + \beta)\dfrac{v_x}{R_B + r_\pi}} = \frac{1}{\dfrac{1}{R_E} + (1 + \beta)\dfrac{1}{R_B + r_\pi}} \tag{8.29}$$

that is

$$r_o = R_E \,//\, \frac{R_B + r_\pi}{1 + \beta} \tag{8.30}$$

Summary: Comparison among the three configurations of transistor amplifiers

(1) The common-emitter amplifiers characterize high voltage and current gain, moderate input resistance, high output resistant but comparably narrow bandwidth. They often appear as input stages, intermediate stages and output stages in low frequency amplifiers. Besides, what to be noticed is that the output voltage is phase-shifted 180° with respect to the input signal.

(2) The common-base amplifiers can only amplify voltages. They characterize voltage gains as high as those of common-emitter amplifiers and the widest bandwidth among the three configurations. Thus they are often used in high frequency designs.

(3) The common-collector amplifiers, aka emitter followers, whose voltage gains are less than but very close to unity, can only amplify currents. Thus they are often used in power amplifiers. The common-collector amplifiers characterize the highest input resistance and lowest output resistance. Thus they are often used as input stages or output stages to do the impedance transformation. Besides, its output and input are in the same phase, leading to a positive voltage gain.

▶ **Problems**

P8.1   As shown in Table P8.1, the voltages at base, emitter and collector of four NPN transistors are given. Suppose for each transistor, we have $V_{BE} = 0.5$ V. Determine the operation regions of the transistors.

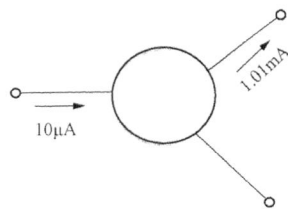

Fig. P8.1   Diagram for P8.3

Table P8.1   **The voltages at base emitter and collector**

Transistor	Unit	VT1	VT2	VT3	VT4
Base Voltage	V	0.7	1	−1.2	0
Emitter Voltage	V	0	0.3	−1.9	0
Collector Voltage	V	3.3	0.7	0	12

P8.2 Two transistors are available for designing a single transistor amplifier with supply voltage of 30V. The parameters of the two candidate transistors are listed in Table P8.2. If the moderate gain is required, determine which transistor or none of them is suitable.

Table P8.2 **The parameters of the two candidate transistors**

Parameters	VT1	VT2
I_{CBQ} （μA）	0.02	0.05
V_{CEQ} （V）	50	20
β	15	100

P8.3 In a specific circuit, the currents are measured and their directions are marked at two terminals of a transistor, as shown in Fig. P8.1. Determine the current of the third terminal and mark its direction. Draw the transistor in the circle and calculate its current gain β.

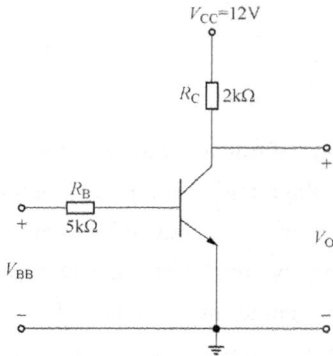

Fig. P8.2 Circuit diagram for P8.4

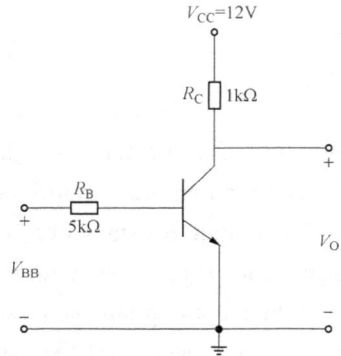

Fig. P8.3 Circuit diagram for P8.5

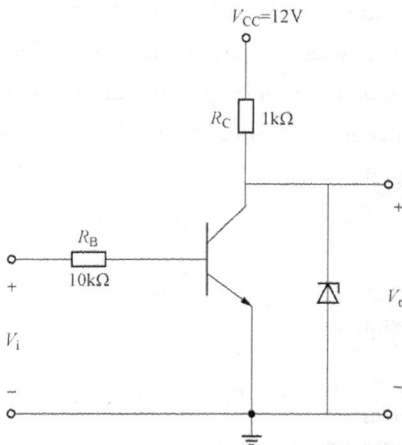

Fig. P8.4 Circuit diagram for P8.6

P8.4 As shown Fig. P8.2, when the transistor is on, $V_{BE}=0.7$ V and $\beta = 50$. Analyze the operation region of the transistor and calculate the output voltage V_O when V_{BB} is 0, 1V, and 3V, respectively.

P8.5 As shown in Fig. P8.3, the power supply V_{BB} is variable. When the transistor is on, $V_{BE}=0.7$V and $\beta = 50$. Determine the voltage of V_{BB} when the transistor enters into saturation region.

P8.6 In the circuit shown in Fig. P8.4, the transistor has $\beta = 50$, $|V_{BE}|=0.2$ V, and $|V_{CES}|=0.1$ V. As for the Zener diode, the Zener voltage is $V_Z=5$ V, and its forward biased voltage is $V_D=0.5$ V. Find the output voltage V_o when $V_i=0$ V and $V_i=5$ V, respectively.

P8.7 For the circuits shown in Fig. P8.5, determine which region each transistor operates in.

P8.8 For the circuit shown in Fig. P8.6, if the bottom of the output waveform is distorted, determine the operation region of the transistor.

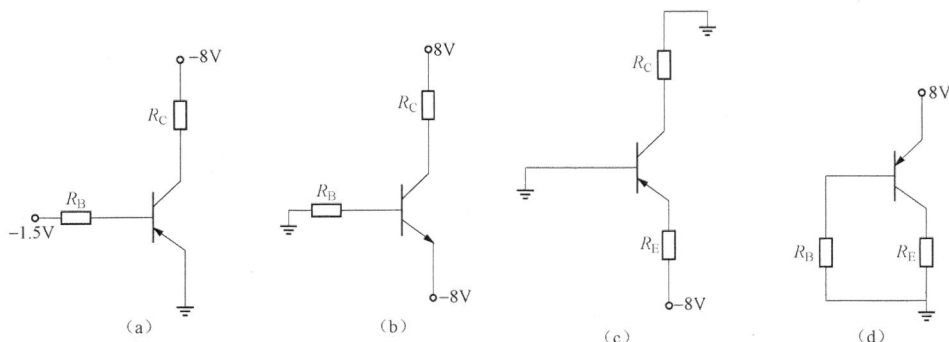

Fig. P8.5   Circuit diagram for P8.7

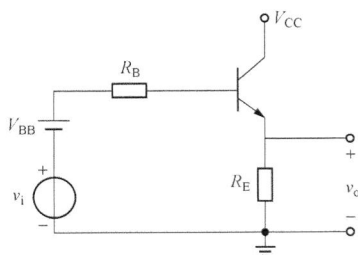

Fig. P8.6   Circuit diagram for P8.8

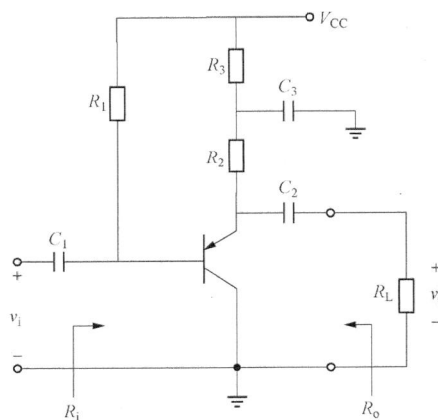

Fig. P8.7   Circuit diagram for P8.10

P8.9   As for the single transistor amplifier, among the three configurations, please select the optimal configuration for voltage amplification, current amplification, and voltage follower, respectively.

P8.10   An amplifier circuit is shown in Fig. P8.7, in which the capacitor acts as a short circuit for ac signal.

(1) Determine the Q point (derive the expressions for I_{BQ}, I_{CQ}, V_{CEQ});

(2) Derive the voltage gain A_v, the input impedance r_i, and the output impedance r_o;

(3) If C_3 is replaced by an open circuit, find the voltage gain A_v, the input impedance r_i, and the output impedance r_o.

场 效 应 管
Field-effect Transistors

场效应管也是很重要的一种半导体器件。物理中"场效应"是指某种材料的电导率可以通过外加电场进行控制的现象，而场效应管正是利用半导体的这一特性，通过控制输入回路的电场效应来控制输出回路电流，达到小信号放大、阻抗变换、恒流源等目的。场效应管广泛应用于放大器及逻辑电路中。

场效应管的种类多种多样，本章首先以最常用的一种——绝缘栅型场效应管为例介绍场效应管的工作原理及其在三个工作区中的伏安特性，然后分别对场效应管的放大电路进行大信号直流分析和小信号交流分析，在此基础上讨论几种常见的场效应管放大电路的实际应用，最后介绍如何用场效应管实现非、与、或等基本逻辑功能。

9.1 绝缘栅型场效应管的工作原理与伏安特性

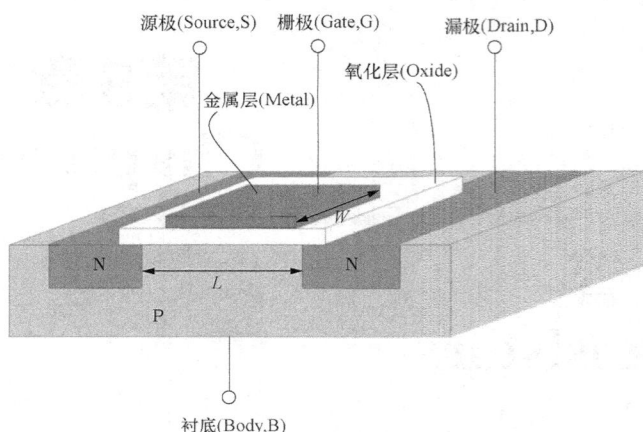

图 9.1 N 通道增强型绝缘栅型场效应管示意图

Fig. 9.1 An illustration of an N-channel enhancement-mode MOSFET

9.1.1 绝缘栅型场效应管的基本工作原理

通过前面的学习，知道半导体材料分 N 型和 P 型两种，通过在硅晶芯片的不同地方添加杂质，可在一块硅晶芯片上同时形成 N 型和 P 型材料以制造晶体管。绝缘栅型场效应管的示意图如图 9.1 所示，它有漏极、栅极、源极和衬底共四个电极，其中栅极由金属材料制成并与衬底绝缘，源极和漏极为 N 型材料，基极为 P 型材料，因此这是一个 N 通道增强型场效应管，简称 NMOS 管。NMOS 管的电路元件符号如图 9.2（a）所示。若绝缘栅型场效应管的源极和漏极由 P 型材料制成而衬底由 N 型材料制成，则称为 PMOS 管，其电路元件符号如图 9.2（b）所示。本章以 NMOS 管为例讨论场效应管的基本工作原理及在三个工作区内的伏安特性，而对 PMOS 管的分析则留给读者完成。

1. 工作在截止区

考虑如图 9.3 所示情况，这里源极与衬底连在一起形成了一个三极器件，为常见接法。假设在漏极和源极之间加有正电压 v_{DS} 而栅极与源极间电压 v_{GS} 为零，此时漏极与衬底之间、源极与衬底之间各形成一个 PN 结。由于 PN 结单向导电性，此时漏极电压高于衬底，因此没有电流流入漏极。这种情况称为场效应管工作于截止区。

接下来逐渐增加 v_{GS} 的值，在 v_{GS} 达到某一阈值 V_T 之前，NMOS 都处于截止区，并具有如下伏安特性

$$i_D = 0, \quad v_{GS} \leqslant V_T \tag{9.1}$$

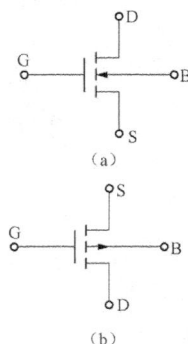

图 9.2 场效应管的电路元件符号
(a) NMOS；(b) PMOS

Fig. 9.2 Circuit symbol for an N-channel enhancement-mode MOSFET
(a) NMOS; (b) PMOS

Field-effect transistors (FETs) are an important type of semiconductor devices. In physics, 'field effect' refers to the modulation of the electrical conductivity of some material by the application of an external electric field. Using this feature, an FET controls the field effect of the input circuit and thus controls the current of the output circuit, so that it can function as an amplifier, current source, etc. FETs are widely used in amplifiers and logic gates.

There are various types of FETs. In this chapter, we discuss the most commonly used metal oxide-semiconductor field-effect transistor (MOSFET). We start from an overview of MOSFETs to understand their operation in three operation regions. Afterwards, we analyze the large-signal DC circuits that bias the device into a proper operation region and use small-signal equivalent circuits to analyze the amplifier. Consequently, some practical applications of FET amplifiers will be discussed. Finally, we use MOSFETs to build basic logic gates such as inverter, NAND gate and NOR gate.

9.1　Overview of MOSFET Operation

9.1.1　MOSFET transistors

In previous chapters, we have learnt that there are two types of semiconducting material, the N-type and the P-type. By adding impurities to different regions of a chip of silicon crystal, we can produce a transistor that has both N-type and P-type material. An illustration of an MOSFET is given in Fig. 9.1. The MOSFET has four terminals, the drain (D), gate (G), source (S), and body (B), where the gate is made from metal and insulated from the body. This is an example of N-channel enhancement-mode MOSFET (shortened as NMOS), where the source and the drain are made of N-type material and the body is made of P-type material. The circuit symbol for an NMOS transistor is shown in Fig. 9.2(a). A MOSFET whose source and drain are made of P-type material whereas whose body is made of N-type material is called a PMOS transistor and its circuit symbol is given in Fig. 9.2(b). In this chapter, we discuss the basic operation of NMOS transistors in three operation regions, while the analysis of PMOS transistors is handed over to readers.

1.　Operation in the cutoff region

Consider the situation shown in Fig. 9.3. Note that in this case, the body is connected to the source to form a three-terminal device, which is quite common in practice. Suppose a positive voltage v_{DS} is applied between the drain and the source, and the voltage between the gate and the source is zero, i.e. $v_{GS}=0$. Therefore, PN junctions appear at the drain/body and at the source/body interfaces. As we know, if positive voltage is added on the P-side, electrons flow easily across a PN junction; whereas if positive voltage is added on the N-side, virtually no current flows. In this case, as shown in Fig. 9.3, positive voltage v_{DS} is added on the N-side of the drain/body junction. Thus, no current flows into the drain terminal. This is called the cutoff region of operation.

式中，i_D 为流入漏极的电流。

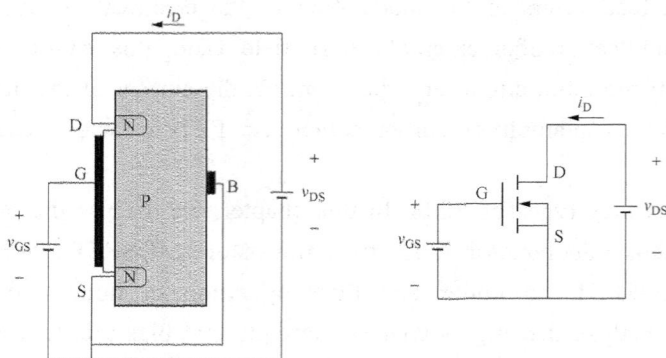

图 9.3 工作在截止区示意图

Fig. 9.3 Operation in the cutoff region

2. 工作在线性区

继续增加栅极电压 v_{GS} 至 $v_{GS} \geqslant V_T$ 并保持 $v_{DS} \leqslant v_{GS} - V_T$，此时 NMOS 处于线性区（也称非饱和区）。如图 9.4 所示，栅极电压所形成的电场在栅极与绝缘层之间感应出正电荷，在绝缘层和 P 型衬底之间感应出负电荷，这些负电荷流向源极，将漏极与源极的 N 型材料连接起来形成导电沟道。当 v_{DS} 增加时，漏极电流流入漏极，经过沟道，从源极流出。

若栅极电压 v_{GS} 固定，当漏极电压 v_{DS} 较小时，漏极电流 i_D 与 v_{DS} 呈正比。可见在线性区，NMOS 管可等效为一个漏极与源极间的电阻，其阻值随 v_{GS} 的增加而减小。

当电压 v_{DS} 较小时，沟道呈矩形。若继续增加 v_{DS}，则随着 v_{DS} 增大，栅极与漏极间电压 v_{GD} 逐渐减小，导致漏极处沟道逐渐变薄，如图 9.5 所示。在这一过程中，漏极电流 i_D 仍随着 v_{DS} 增加而增加，但增长速度变慢，二者之间不再呈线性关系。

当 $v_{GS} \geqslant V_T$ 且 $v_{DS} \leqslant v_{GS} - V_T$ 时，NMOS 工作于线性区，漏极电流为

$$i_D = K[2(v_{GS} - V_T)v_{DS} - v_{DS}^2] \tag{9.2}$$

式中，系数 K 为

$$K = \frac{W}{L}\frac{KP}{2} \tag{9.3}$$

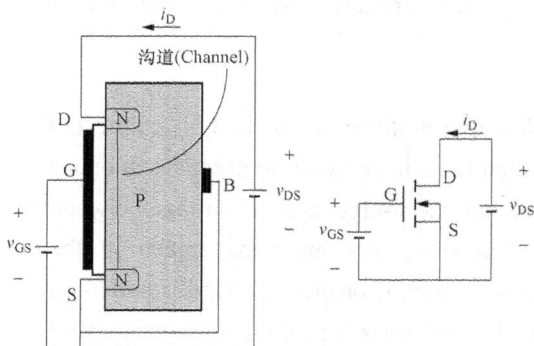

图 9.4 工作在线性区示意图

Fig. 9.4 Operation in the triode region

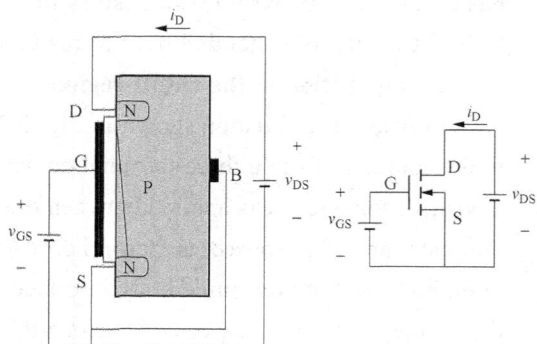

图 9.5 v_{DS} 继续增加，漏极处沟道变薄

Fig. 9.5 As v_{DS} increases, the channel pinches down at the drain end

As v_{GS} is increased, the NMOS remains in cutoff until v_{GS} reaches a particular value called the threshold voltage V_T. Thus in the cutoff region, we have

$$i_D = 0, \quad v_{GS} \leqslant V_T \tag{9.1}$$

where i_D is the current flowing into the drain terminal.

2. Operation in the triode region

Keeping increasing the gate voltage v_{GS} until $v_{GS} \geqslant V_T$ and making sure $v_{DS} \leqslant v_{GS} - V_T$, the device operates in the triode region. (The triode region is also called the linear region of operation.) In this case, as shown in Fig. 9.4, the electric field resulted from v_{GS} produces holes between the gate and the insulator, and electrons between the insulator and the body. The electrons flow towards the source, which results in an N-type channel between the drain and the source. Then, when v_{DS} is increased, current flows into the drain, through the channel, and out of the source.

For small values of v_{DS}, when gate voltage v_{GS} is fixed, drain current i_D is proportional to v_{DS}. In the triode region, the NMOS behaves as a resistor connected between the drain and the source, and its resistance decreases as v_{GS} increases.

When v_{DS} is small, the channel is rectangle. If v_{DS} is increased, the voltage between the gate and the drain decreases, resulting in tapering of the channel thickness towards the drain end, as illustrated in Fig. 9.5. During this procedure, the drain current i_D increases as v_{DS} increases. However, it increases more slowly and the relationship between i_D and v_{DS} is no longer linear.

In the triode region, the drain current is given by

$$i_D = K \left[2(v_{GS} - V_T)v_{DS} - v_{DS}^2 \right] \tag{9.2}$$

where K is obtained from

$$K = \frac{W}{L} \frac{KP}{2} \tag{9.3}$$

in which W is the width of the channel and L is the length, as illustrated in Fig. 9.1, and KP is a device parameter that depends on the thickness of the oxide layer and certain properties of the channel material. A typical value of KP for N-channel enhancement devices is $50\,\mu A / V^2$.

It should be noted that as $v_{GD} = v_{GS} - v_{DS}$, the condition $v_{DS} \leqslant v_{GS} - V_T$ is equivalent to $v_{GD} \geqslant V_T$.

More facts about transistors

Bipolar junction transistors (BJTs) are widely used in electronic equipment, including computers, televisions, mobile phones, audio amplifiers, industrial control, and radio transmitters. They are available as individual components, or fabricated in integrated circuits, often in large numbers. BJTs use both electron and hole charge carriers. Field-effect transistors (FETs), on the other hand, only use one type of charge carrier in a semiconductor material, and are also known as unipolar transistors. A FET typically produces less noise than a BJT, and typically has better thermal stability than a BJT. However, it has a relatively low gain-bandwidth product compared to a BJT.

这里 W 和 L 分别表示沟道的宽与长（见图 9.1），KP 是由材料特性所决定的元件参数，其值通常为 $50\mu A/V^2$。

考虑到 $v_{GD} = v_{GS} - v_{DS}$，NMOS 场效应晶体管工作于线性区的条件之一 $v_{DS} \leqslant v_{GS} - V_T$ 等价于 $v_{GD} \geqslant V_T$。

3. 工作在饱和区

如前所述，当 v_{DS} 增加时，漏极处的沟道厚度逐渐变薄。当达到 $v_{DS} = v_{GS} - V_T$ 时，在漏极处沟道厚度为零，在源极处沟道仍维持原先的厚度，沟道变成楔形。此时 $v_{GD} = V_T$，场效应管处于临界状态。然后继续增加 v_{DS}，但 i_D 保持不变，如图 9.6 所示，这一状态称为饱和区。NMOS 工作于饱和区的条件为 $v_{GS} \geqslant V_T$ 且 $v_{DS} \geqslant v_{GS} - V_T$，漏极电流 i_D 为

$$i_D = K(v_{GS} - V_T)^2 \tag{9.4}$$

4. NMOS 管小结

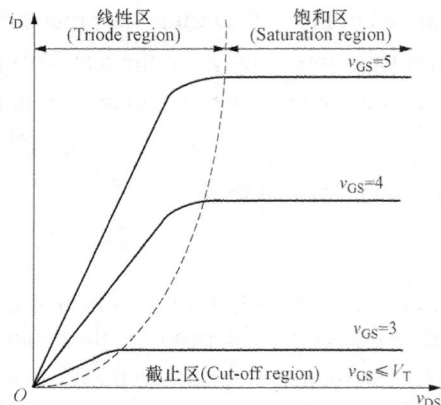

图 9.7 所示为 NMOS 管漏极伏安特性曲线。由图可知，当 v_{GS} 小于阈值时，NMOS 管处于截止区。当 v_{GS} 大于阈值时，NMOS 管工作于线性区或饱和区；对应不同的 v_{GS} 值，场效应管具有不同的伏安特性曲线。当处于线性区与饱和区边界时，即当 $v_{DS} = v_{GS} - V_T$ 时，有 $i_D = Kv_{DS}^2$，即线性区与饱和区的边界为抛物线。显然，场效应管是一种由电压控制电流的元件。

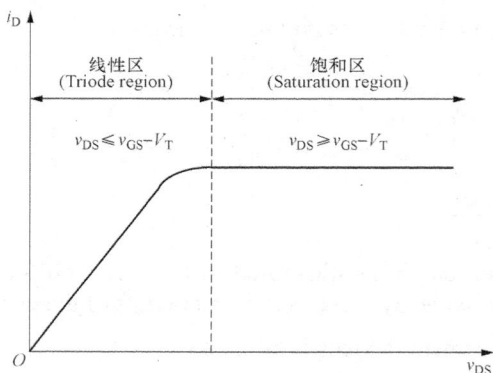

图 9.6 v_{DS} 增加至 $v_{DS} \geqslant v_{DS} - v_T$ 后，i_D 保持不变
Fig.8.6 As v_{DS} increases to $v_{DS} \geqslant v_{GS} - V_T$, i_D becomes constant

图 9.7 NMOS 管漏极伏安特性曲线
Fig.9.7 Characteristic curves for an NMOS transistor

当 NMOS 管的栅极与源极之间加了足够大的正电压后，电子被吸引到栅极下方的绝缘层与衬底之间，从而在漏极与源极之间形成一条 N 型导电沟道。此时在漏极与源极间加正电压，则有电流从漏极流入，流经沟道，再从源极流出。漏极电流的大小由栅极与源极间电压所控制。

9.1.2 PMOS 管

PMOS 管的构造与 NMOS 管刚好相反，其漏极和源极由 P 型材料制成，而衬底为 N 型材料。两种场效应管的特性非常相似，除了电压方向相反。在 PMOS 中，以电流流出漏极的方向为正。当栅极与源极间电压为负且大于阈值（阈值为负数）时，PMOS 处于截止区。继续

3. Operation in the saturation region

As mentioned before, the thickness of the channel at the drain end becomes thinner as v_{DS} increases. When v_{DS} is increased to $v_{DS}=v_{GS}-V_T$, the channel thickness at the drain end becomes zero, while that at the source end remains at its original value. Hence, the channel becomes wedge. In this case, we have $v_{GD}=V_T$ and the device works at the boundary. Further increasing v_{DS}, i_D remains constant, as illustrated in Fig. 9.6. This is called the saturation region, in which we have $v_{GS}\geqslant V_T$ and $v_{DS}\geqslant v_{GS}-V_T$, and the drain current is given by

$$i_D=K(v_{GS}-V_T)^2 \tag{9.4}$$

4. Summary of NMOS transistors

The characteristic curves for an NMOS transistor are illustrated in Fig. 9.7 As can be seen from the figure, in the cutoff region, v_{GS} is smaller than the threshold. In the triode and saturation regions, v_{GS} is greater than the threshold; for various v_{GS}, the curves are different. At the boundary between the triode region and the saturation region, i.e. when $v_{DS}=v_{GS}-V_T$, we have $i_D=Kv_{DS}^2$. The boundary between the triode region and the saturation region is a parabola. Apparently, MOSFETs are a kind of device whose current is controlled by voltage.

> For an NMOS transistor, when a sufficiently large (positive) voltage is applied between the gate and the source, electrons are attracted to the region under the gate, between the insulator and the body. Hence, a channel of N-type material is induced between the drain and the source. Then, if positive voltage is applied to the drain relative to the source, current flows into the drain through the channel and out of the source. Drain current is controlled by the voltage applied to the gate.

9.1.2   PMOS transistors

Interchanging the N and P regions of N-channel devices, we have P-channel devices. PMOS transistors have P-type material for the drain and source terminals and N-type material for the body. The characteristics of a PMOS transistor are very similar to those of an NMOS transistor, except that voltage polarities are inverted. For a PMOS transistor, the positive direction of the drain current is referenced as flowing out of the drain. When the voltage between the gate and the source, v_{GS}, is negative and is greater than a threshold (the threshold value is also negative), the PMOS transistor works in the cutoff region. Continuing to decrease v_{GS}, electrons are attracted to the gate, forming a P-type material channel under the gate and between the source and the drain. Then, if negative voltage is applied to the drain relative to the source, current flows into the source, through the channel, and out of the drain. A summary of the characteristics of NMOS and PMOS transistors is given in Table 9.1.

减小 v_{GS}（$v_{GS} < 0$），电子被吸引至栅极，从而在栅极下方形成 P 型沟道，该沟道连接源极与漏极。此时在漏极与源极间加足够大的负电压，则有电流从源极流入，流经沟道，从漏极流出。NMOS 与 PMO 在三个工作区的伏安特性总结在表 9.1 中。

表 9.1 　　　　　　　　　　　　场效应管伏安特性小结
Table 9.1 　　　　　　　　　　　MOSFET Summary

项目（Items）	NMOS	PMOS
电路元件符号（Circuit symbol）		
V_T（典型值）［V_T (typical value)］	$V_T = 1\,\text{V}$	$V_T = -1\,\text{V}$
截止区（Cut-off region）	$v_{GS} \leq V_T$ $i_D = 0$	$v_{GS} \geq V_T$ $i_D = 0$
线性区（Triode region）	$v_{GS} \geq V_T$ 且（and）$0 \leq v_{DS} \leq v_{GS} - V_T$ $i_D = K[2(v_{GS} - V_T)v_{DS} - v_{DS}{}^2]$	$v_{GS} \leq V_T$ 且（and）$0 \geq v_{DS} \geq v_{GS} - V_T$ $i_D = K[2(v_{GS} - V_T)v_{DS} - v_{DS}{}^2]$
饱和区（Saturation region）	$v_{GS} \geq V_T$ 且（and）$v_{DS} \geq v_{GS} - V_T$ $i_D = K(v_{GS} - V_T)^2$	$v_{GS} \leq V_T$ 且（and）$v_{DS} \leq v_{GS} - V_T$ $i_D = K(v_{GS} - V_T)^2$
v_{GS} 及 v_{DS}	通常设为正值 Normally assume positive values	通常设为负值 Normally assume negtive values

9.2 场效应管放大电路的直流大信号分析

在应用场效应管构造功能电路之前，首先要确定它的工作区，使其特性满足所要构造的功能电路的要求。这里以最常用的放大电路为例进行分析。首先根据直流大信号电路，分析场效应管工作在哪个工作区，即确定静态工作点。在此基础上，根据交流小信号等效电路计算输入阻抗、放大倍数等。本节讨论第一步，静态工作点的确定。

图 9.8（a）所示为常用于确定静态工作点的偏置电路，其中 V_{DD} 为用于偏置场效应管的直流电压源。之所以称之为大信号，是因为与被放大的毫伏级小信号相比，V_{DD} 通常为伏级。根据上一节所述内容，为确定场效应管的工作区，需要计算漏极电流 i_D 与栅源极电压 v_{GS}。

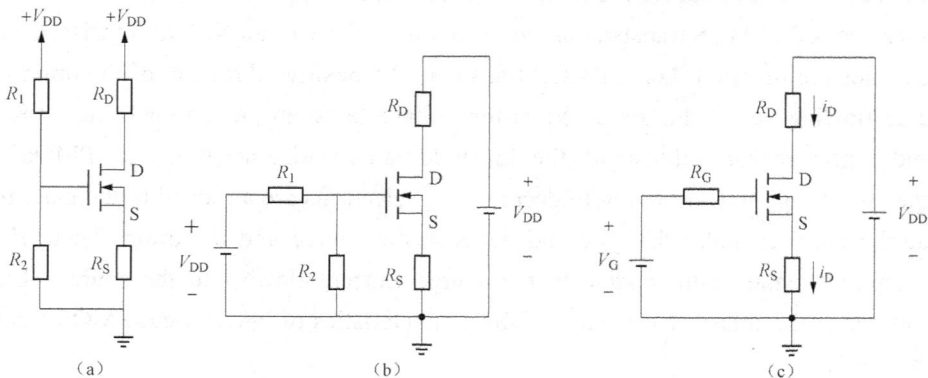

图 9.8 场效应管直流大信号偏置电路

Fig. 9.8 The bias circuit for an Nmos transistor

9.2 Large–signal DC Analysis of NMOS Transistors

To design an MOSFET circuit for a certain purpose, the first task would be to determine which region it should work in, so that the desired functions can be achieved. In this chapter, we discuss the most common application——amplifier circuits. First, we analyze the large-signal DC circuit to determine which region the MOSFET operates in, i.e. to determine the Q point. Then, we use a small-signal equivalent circuit to determine the input resistance, voltage gain, etc. In this section, we discuss the first step.

A common bias circuit used for establishing the Q point is shown in Fig. 9.8(a), where V_{DD} is the DC voltage source used to bias the MOSFET. It is called the 'large' signal because V_{DD} usually has a value of several volts, compared with the 'small' signal of several milli-volts that is to be amplified. According to the previous section, in order to determine the transistor's operation region, we need to calculate the drain current i_D and the voltage between the gate and the source v_{GS}.

More facts about transistors

A point-contact transistor was the first type of solid-state electronic transistor ever constructed. It was developed by research scientists John Bardeen and Walter Brattain at Bell Laboratories in 1947. They worked in a group led by physicist William Shockley. The group had been working together on experiments and theories of electric field effects in solid state materials, with the aim of replacing vacuum tubes with a smaller, less power-consuming device.

The critical experiment consisted of a block of germanium, a semiconductor, with two very closely spaced gold contacts held against it by a spring. Brattain attached a small strip of gold foil over the point of a plastic triangle -- a configuration which is essentially a point-contact diode. He then carefully sliced through the gold at the tip of the triangle, which produced two electrically isolated gold contacts very close to each other. This resulted in that a small positive current applied to one of the two contacts had an influence on the current which flowed between the other contact and the base upon which the block of germanium was mounted.

The point-contact transistor was commercialized and sold by Western Electric and others but was soon superseded by the bipolar junction transistor, which was easier to manufacture and more rugged. Germanium was employed extensively for two decades in the manufacture of transistors, but was then almost totally replaced by silicon and other alloyed materials.

为方便分析，首先将原电路转化为如图 9.8（b）所示形式。然后利用戴维南等效定理化简左边电路，如图 9.8（c）所示。此时戴维南电压为

$$V_{\mathrm{G}} = V_{\mathrm{DD}} \frac{R_2}{R_1 + R_2}$$

戴维南电阻为

$$R_{\mathrm{G}} = \frac{R_1 R_2}{R_1 + R_2}$$

由于流入漏极的电流与流出源极的电流相同，栅极几乎没有电流通过，所以电阻 R_{G} 上的压降几乎为零，因此对栅极回路应用 KVL 可得

$$V_{\mathrm{G}} = v_{\mathrm{GS}} + R_{\mathrm{S}} i_{\mathrm{D}} \tag{9.5}$$

为了使场效应管起到放大器的作用，需要将其偏置至饱和区，此时

$$i_{\mathrm{D}} = K(v_{\mathrm{GS}} - V_{\mathrm{T}})^2 \tag{9.6}$$

联立求解式（9.5）与式（9.6），即可得到静态工作点。两个方程的曲线如图 9.9 所示，其中由式（9.5）得到的直线称为偏置线。可以看出两条曲线共有两个交点，即联立两式得到的方程组有两组根，其中 v_{GS} 较小的那个根不满足 $v_{\mathrm{GS}} > V_{\mathrm{T}}$ 这一条件，应当舍去。因此，最终的静态工作点确定为较大的 v_{GS} 与较小的 i_{D} 这组根，并将其值相应地记为 V_{GSQ} 与 I_{DQ}。

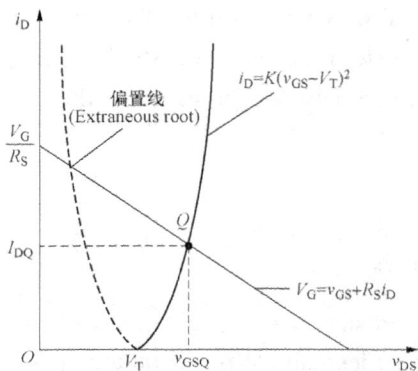

图 9.9　式（9.5）与式（9.6）的曲线

Fig. 9.9　Graphical solution of (9.5) and (9.6)

最后，如图 9.8（c）所示，对漏极回路应用 KVL 得

$$V_{\mathrm{DSQ}} = V_{\mathrm{DD}} - (R_{\mathrm{D}} + R_{\mathrm{S}}) I_{\mathrm{DQ}} \tag{9.7}$$

9.3　场效应管放大电路的小信号等效电路

在场效应管被偏置到饱和区后，可通过小信号等效电路分析其放大功能。小信号是需要被场效应管放大的输入信号，由于其幅度很小，它与直流大信号的叠加并不会对静态工作点产生影响，场效应管仍工作在饱和区。有小信号输入后，漏极电流由两部分组成，即

$$i_{\mathrm{D}}(t) = I_{\mathrm{DQ}} + i_{\mathrm{d}}(t) \tag{9.8}$$

式中，I_{DQ} 为静态工作点的直流大信号电流；$i_{\mathrm{d}}(t)$ 为输入小信号引起的电流。

同样，栅源极电压也由静态直流信号与输入交流小信号两部分组成，即

$$v_{\mathrm{GS}}(t) = V_{\mathrm{GSQ}} + v_{\mathrm{gs}}(t) \tag{9.9}$$

将式（9.8）与式（9.9）代入 $i_{\mathrm{D}} = K(v_{\mathrm{GS}} - V_{\mathrm{T}})^2$，可得

$$I_{\mathrm{DQ}} + i_{\mathrm{d}}(t) = K[V_{\mathrm{GSQ}} + v_{\mathrm{gs}}(t) - V_{\mathrm{T}}]^2 \tag{9.10}$$

将上式右端展开，有

$$I_{\mathrm{DQ}} + i_{\mathrm{d}}(t) = K(V_{\mathrm{GSQ}} - V_{\mathrm{T}})^2 + 2K(V_{\mathrm{GSQ}} - V_{\mathrm{T}})v_{\mathrm{gs}}(t) + Kv_{\mathrm{gs}}^2(t) \tag{9.11}$$

For the convenience of analysis, the original circuit is redrawn as shown in Fig. 9.8(b). Replacing the gate circuit with its Thévenin equivalent, we have a simplified circuit as shown in Fig. 9.8 (c), in which the Thévenin voltage is $V_G = V_{DD} \dfrac{R_2}{R_1 + R_2}$ and the Thévenin resistance is $R_G = \dfrac{R_1 R_2}{R_1 + R_2}$. As the current flows into the drain and out of the source and almost no current flows through the gate, the voltage drop across R_G is considered to be zero. Therefore, applying KVL to the gate loop, we have

$$V_G = v_{GS} + R_S i_D \qquad (9.5)$$

In order to obtain an amplifier, the transistor should be biased to the saturation region, so we have

$$i_D = K(v_{GS} - V_T)^2 \qquad (9.6)$$

We can get the operating point by solving for (9.5) and (9.6) simultaneously. The plots of the two equations are given in Fig. 9.9. The straight line of (9.5) is called the bias line. As can be seen from the figure, the two plots have two intersection points, i.e. the system of equations has two groups of roots, among which the smaller root found for v_{GS} does not satisfy the condition of $v_{GS} > V_T$, and thus should be discarded. The larger root found for v_{GS} and the smaller root found for i_D are the true operating point. In this case, denote them by V_{GSQ} and I_{DQ}, respectively.

Finally, applying KVL to the drain loop, as shown in Fig. 9.8(c), we have

$$V_{DSQ} = V_{DD} - (R_D + R_S)I_{DQ} \qquad (9.7)$$

9.3   Small–signal Equivalent Circuits of NMOS Transistors

Since the NMOS transistor is biased into the saturation region, we use small-signal equivalent circuits to analyze its amplification function. The small signal is the signal to be amplified; as it has very low amplitude, it only results small changes from the Q point. Thus, the MOSFET still operates in the saturation region. The drain current thereby consists of two parts

$$i_D(t) = I_{DQ} + i_d(t) \qquad (9.8)$$

where I_{DQ} is the DC large signal at the Q point and $i_d(t)$ is resulted from the small signal input into the MOSFET. Similarly, the voltage between the gate and the source also consists of two parts

$$v_{GS}(t) = V_{GSQ} + v_{gs}(t) \qquad (9.9)$$

Substituting (9.8) and (9.9) into $i_D = K(v_{GS} - V_T)^2$, we have

$$I_{DQ} + i_d(t) = K[V_{GSQ} + v_{gs}(t) - V_T]^2 \qquad (9.10)$$

The right-hand side of (9.10) can be expanded to obtain

$$I_{DQ} + i_d(t) = K(V_{GSQ} - V_T)^2 + 2K(V_{GSQ} - V_T)v_{gs}(t) + Kv_{gs}^2(t) \qquad (9.11)$$

由于

$$I_{DQ} = K(V_{GSQ} - V_T)^2 \tag{9.12}$$

上式左右两端第一项可消掉。由于与 $\left| V_{GSQ} - V_T \right|$ 相比 $\left| v_{gs}(t) \right|$ 非常小，式（9.11）右端第三项可忽略不计。因此式（9.11）变为

$$i_d(t) = 2K(V_{GSQ} - V_T)v_{gs}(t) \tag{9.13}$$

定义

$$g_m = 2K(V_{GSQ} - V_T) \tag{9.14}$$

为场效应管的跨导，则式（9.13）可简写为

$$i_d(t) = g_m v_{gs}(t) \tag{9.15}$$

考虑到栅极电流可忽略，即 $i_g(t) = 0$，场效应管小信号等效电路可由图 9.10 表示。这里，场效应管用一个电压控制电流源等效，且其栅极与源极间开路。

场效应管的跨导 g_m 是一个重要参数，一般来说 g_m 越大，场效应管性能越好。g_m 由静态工作点和场效应管参数决定。式（9.12）和式（9.14）可得

$$g_m = 2\sqrt{KI_{DQ}} \tag{9.16}$$

将式（9.3）参数 K 的定义代入式（9.16），可得

$$g_m = \sqrt{2KP}\sqrt{W/L}\sqrt{I_{DQ}} \tag{9.17}$$

在上述场等效电路中，饱和区 i_D 假设为一条水平直线，然而实际中 i_D 是随着 v_{DS} 的增加而极缓慢地增加的。如果需要在等效电路中体现这一特性，则需要在漏极与源极之间加一个电阻 r_d，该电阻称为漏极电阻。考虑漏极电阻的等效电路如图 9.11 所示，此时式（9.15）变为

$$i_d(t) = g_m v_{gs}(t) + v_{ds}(t)/r_d \tag{9.18}$$

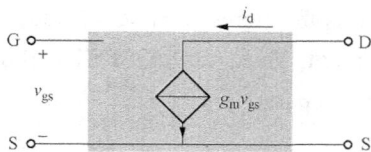

图 9.10　场效应管小信号等效电路
Fig. 9.10　Small-signal equivalent circuit for an MOSFET

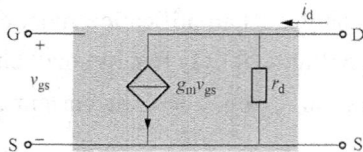

图 9.11　考虑漏极电阻的场效应管小信号等效电路
Fig. 9.11　Small-signal equivalent circuit that involves the drain resistor

讨论：　几种常见的场效应管放大电路的实际应用

1. 共源放大器

共源放大器的电路图如图 9.12 所示，请读者：

（1）从该电路图中找出偏置电路，注意电容的作用。

（2）对偏置电路进行直流大信号分析，确定静态工作点。

（3）根据该电路图画出小信号等效电路，计算其放大倍数、输入阻抗和输出阻抗。

As

$$I_{DQ} = K(V_{GSQ} - V_T)^2 \qquad (9.12)$$

the first term on each side of the above equation can be canceled. Thus, (9.11) becomes

$$i_d(t) = 2K(V_{GSQ} - V_T)v_{gs}(t) \qquad (9.13)$$

Defining

$$g_m = 2K(V_{GSQ} - V_T) \qquad (9.14)$$

as the transconductance of the MOSFET, (9.13) can be written as

$$i_d(t) = g_m v_{gs}(t) \qquad (9.15)$$

The gate current is negligible, so we have $i_g(t) = 0$. Therefore, the small-signal equivalent circuit can be illustrated by Fig. 9.10, where the MOSFET is modeled by a voltage-controlled current source, and has an open circuit between the gate and the source.

The transconductance g_m is an important parameter. In general, higher values of g_m result in better performance of MOSFET. The value of g_m is determined by the Q point and device parameters. From (9.12) and (9.14) we have

$$g_m = 2\sqrt{KI_{DQ}} \qquad (9.16)$$

If we use (9.3) to substitute for K in the above equation, we obtain

$$g_m = \sqrt{2KP}\sqrt{W/L}\sqrt{I_{DQ}} \qquad (9.17)$$

In the equivalent circuit mentioned above we have assumed that in the saturation region, the curve of i_D is a horizontal line. However, in practice i_D increases slowly as v_{DS} increases. If we want to involve this characteristic in small-signal equivalent circuit, a resistor r_d must be added between the drain and the source, as shown in Fig. 9.11. Such a resistor is called the drain resistor. In this case, (9.15) becomes

$$i_d(t) = g_m v_{gs}(t) + v_{ds}(t)/r_d \qquad (9.18)$$

Discussion: Practical applications of MOSFET amplifier circuits

1. Common-source amplifiers

The circuit of a common-source amplifier is given in Fig. 9.12. Please:

(1) Find out the bias circuit from the figure, and have a discussion on what is the function of the capacitors;

(2) Analyze the large-signal DC circuit and determine the Q point;

(3) Draw the small-signal equivalent circuit and calculate its gain, input impedance and output impedance.

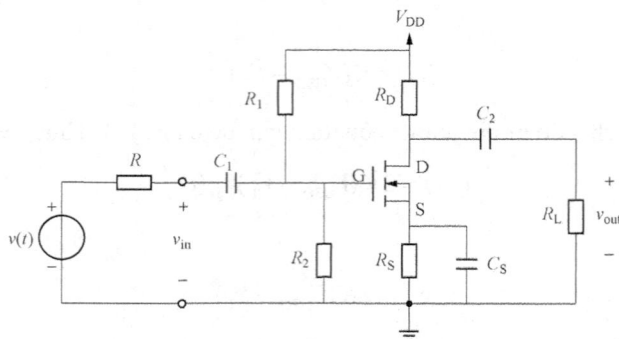

图 9.12　共源放大器

Fig. 9.12　Common-source amplifier

分析：

（1）对于直流情况，电容相当于开路；而对于交流情况，大电容相当于短路。因此由于电容的存在，使得直流偏置电压 V_{DD} 仅对电路中间的场效应管及其周围的四个电阻起作用。因此偏置电路与图 9.8（a）相同，对静态工作点的计算也与第 9.2 节一致。

（2）对于交流小信号而言，电容相当于短路，则有源极直接接地，这也是为什么称之为共源放大器。直流偏置电源相对于交流信号而言也相当于短路，则电阻 R_1 和 R_D 的上端相当于接地。再用场效应管的小信号模型（见图 9.10 或图 9.11）替代，得到如图 9.13 所示的共源放大器小信号等效电路。

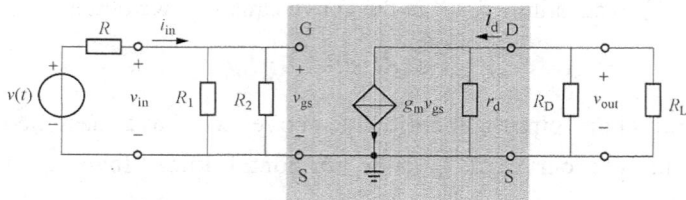

图 9.13　共源放大器的小信号等效电路

Fig. 9.13　The small-signal equivalent circuit common-source amplifer

（3）计算该放大电路的放大倍数即确定输出电压与输入电压的比值。分析右侧电路可知，输出电压为受控电流源的电流与三个并联电阻的等效电阻之积，即

$$v_{out} = -(g_m v_{gs})R_L'$$

式中，$R_L' = \dfrac{1}{1/r_d + 1/R_D + 1/R_L}$，输入电压为 v_{gs}，则放大倍数为

$$A_v = \frac{v_{out}}{v_{in}} = -g_m R_L'$$

（4）放大电路的输入阻抗为电阻 R_1 和 R_2 的并联，即

$$R_{in} = \frac{v_{in}}{i_{in}} = \frac{R_1 R_2}{R_1 + R_2}$$

Analysis:

(1) The capacitors behave as open circuits for DC cases, whereas with large capacitance they are considered as short circuits for AC signals. Therefore, the capacitors make sure that the bias voltage V_{DD} is only added to the NMOS transistor and the four resistors connected to it. Therefore, the bias circuit is the same as the one shown in Fig. 9.8(a), the determination of the Q point is the same as we did in Section 9.2.

(2) For the AC signal, capacitors behave as short circuits; thus, the source is grounded. This is why it is called common-source amplifier. The DC supply voltage source is also considered as a short circuit for the AC signal; hence, the top sides of resistors R_1 and R_D are grounded. Substituting the MOSFET by its small-signal equivalent circuit (as shown in Fig. 9.10 or Fig. 9.11), the small-signal equivalent circuit for the common-source amplifier is shown in Fig. 9.13.

(3) To find the voltage gain of the amplifier circuit, we need to calculate the ratio of output voltage against input voltage. From the circuit on the righthand side we can find that the output voltage is the product of the current from the controlled current source and the equivalent resistance of the three resistors connected in parallel

$$v_{out} = -(g_m v_{gs})R_L'$$

where $R_L' = \dfrac{1}{1/r_d + 1/R_D + 1/R_L}$, and the input voltage is v_{gs}. Consequently, the voltage gain is calculated from

$$A_v = \frac{v_{out}}{v_{in}} = -g_m R_L'$$

(4) The input impedance of the amplifier circuit is the parallel combination of R_1 and R_2

$$R_{in} = \frac{v_{in}}{i_{in}} = \frac{R_1 R_2}{R_1 + R_2}$$

More facts about amplifiers

The input impedance and output impedance are important parameters in the design of a transistor amplifier. An amplifier's impedance values are particularly important for analysis especially when cascading individual amplifier stages together one after another to minimize distortion of the signal. Amplifiers are supposed to have high input impedance, low output impedance, and virtually any arbitrary gain. If an amplifier's input impedance is too low, it can have an adverse loading effect on the previous stage and possibly affecting the frequency response and output signal level of that stage. Some types of amplifier designs, such as the common collector amplifier, automatically have high input impedance and low output impedance by the very nature of their design.

输出阻抗的求解较为复杂，首先将独立源置零，保留受控源，如图 9.14 所示。由于此时 $v_{gs}=0$ ，则受控电流源的输出电流也为零，相当于开路，则输出阻抗为电阻 r_d 和 R_D 的并联，即

$$R_{out} = \frac{r_d R_D}{r_d + R_D}$$

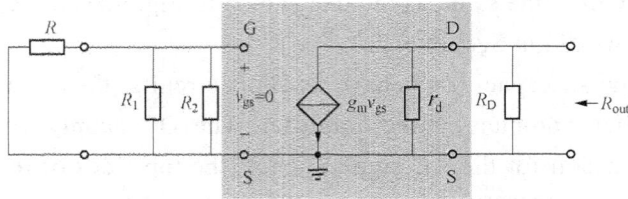

图 9.14　求解共源放大器输出阻抗的电路图

Fig. 9.14　Circuit used for calculating the output impedance of common-source amplifier

2. 源极跟随器

源极跟随器是另一种常见的场效应管放大器，其电路如图 9.15 所示。请读者分析其偏置电路，画出小信号等效电路，并计算电压放大倍数、输入阻抗及输出阻抗。

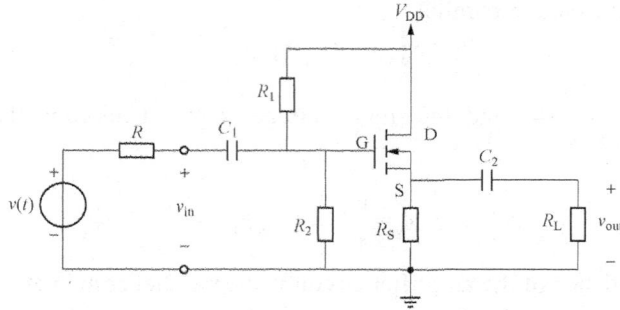

图 9.15　源极跟随器

Fig. 9.15　Source follower

与分析共源放大器相同，在分析源极跟随器的小信号等效电路时，也将电容和偏置电压源视为短路，再用场效应管的小信号等效电路替代。源极跟随器的小信号等效电路如图 9.16 所示。

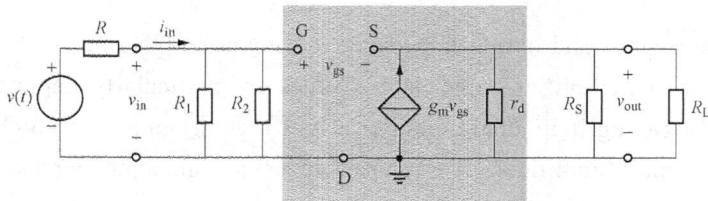

图 9.16　源极跟随器的小信号等效电路

Fig. 9.16　The small-signal equivalent circuit of source follower

输出电压可通过受控电流源的电流与三只并联电阻的等效电阻之积计算出来

$$v_{out} = \frac{g_m v_{gs}}{1/r_d + 1/R_S + 1/R_L} = g_m v_{gs} R'_L$$

The calculation of output impedance is more complex. First we need to zero the independent source and keep the controlled source, as shown in Fig. 9.14 As $v_{gs}=0$, the current of the controlled source is also zero and hence acts as an open circuit. Therefore, the output impedance is the parallel combination of r_d and R_D

$$R_{out} = \frac{r_d R_D}{r_d + R_D}$$

2. Source follower

Source follower is another commonly used MOSFET amplifier, as shown in Fig. 9.15. Please find out its bias circuit, small-signal equivalent circuit, voltage gain, input impedance and output impedance.

The analysis of the small-signal equivalent circuit of source follower is the same as that of the common-source amplifier. We replace the capacitors and the DC supply voltage source by short circuits and substitute the MOSFET by its small-signal equivalent circuit. Consequently, the small-signal equivalent circuit of source follower is given in Fig. 9.16.

The output voltage can be calculated from the product of the current of the controlled current source and the equivalent resistance of the three resistors connected in parallel

$$v_{out} = \frac{g_m v_{gs}}{1/r_d + 1/R_S + 1/R_L} = g_m v_{gs} R_L'$$

More facts about amplifier circuits

Circuit analysis of amplifier circuits can be explained using the following illustration. The BJT or MOSFET transistor is the core element of an amplifier circuit, which is analogous to a TV set. To watch a TV program, one must ensure two things happen: the TV set is turned on and there is TV program on air, so that the TV set is able to receive program and display it. To bias a transistor to the proper operational region (for a BJT it is the active region and for a MOSFET it is the saturation region) is analogous to turn on the TV set, which means the transistor is ready to perform as an amplifier. This is achieved by large-signal DC analysis, where we check based on the configuration of the circuit (including the value of the supply voltage, the resistances of the resistors, the parameters of the transistor, etc.), whether the transistor has been biased into the proper operational region. The TV set receiving program corresponds to the transistor amplifying a small signal, which is achieved by small-signal AC analysis. Based on the small-signal equivalent circuit, we calculate the voltage gain, current gain, input resistance, output resistance, etc., to evaluate the amplifying behavior of the amplifier circuit.

而对于输入电压，有 $v_{in} = v_{gs} + v_{out}$。因此电压放大倍数为

$$A_v = \frac{v_{out}}{v_{in}} = \frac{g_m R_L'}{1 + g_m R_L'}$$

输入阻抗为

$$R_{in} = \frac{v_{in}}{i_{in}} = \frac{R_1 R_2}{R_1 + R_2}$$

输出阻抗的求解方法与前述相同，将独立电压源置零并保留受控电流源，求解从右端看进去的等效电阻。注意此时 $v_{gs} \neq 0$，为求解输出阻抗，可在输出端加一电压源 v_x，并设其输出电流为 i_x，如图 9.17 所示，则输出阻抗为

$$R_{out} = \frac{v_x}{i_x}$$

分析图 9.17 所示电路可得

$$\begin{cases} v_x = \dfrac{i_x + g_m v_{gs}}{1/r_d + 1/R_S} \\ v_x = -v_{gs} \end{cases}$$

从而有

$$R_{out} = \frac{1}{1/r_d + 1/R_S + g_m}$$

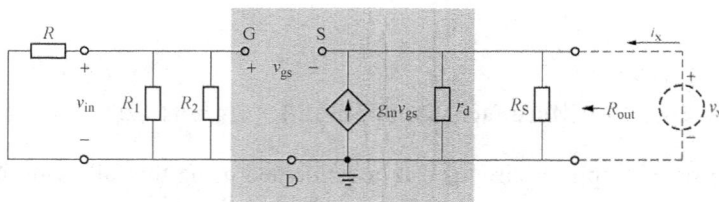

图 9.17 求解源极跟随器输出阻抗的电路图

Fig.9.17 Circuit used for calculating the output impedance of source follower

9.4 用场效应管实现基本逻辑功能

场效应管的一个重要应用是制造各种逻辑电路。本节介绍如何用场效应晶体管构造基本的逻辑门电路。

9.4.1 非门

非门电路图如图 9.18（a）所示，其中包含一个 PMOS 和一个 NMOS，电源电压 V_{DD} 加在 PMOS 的源极。当输入电压 V_{in} 较大时（$V_{in} = V_{DD}$），PMOS 处于截止区，相当于开路；而 NMOS 的漏极与源极间出现导电沟道，相当于一阻值较小的电阻，可用短路等效。此时图 9.18（a）的等效电路如图 9.18（b）所示，电路的输出电压 $V_{out} = 0$。当输入电压 V_{in} 较小时，PMOS 的栅极下方出现导电沟道而 NMOS 处于截止区，因此 PMOS 导通而 NMOS 断开，其等效电路如图 9.18（c）所示，输出电压 $V_{out} = V_{DD}$。

由于场效应管的开关效应，当输入电压较高时，输出电压较低；反之亦然。该电路完全实现了非门的功能。

For input voltage, we have $v_{in} = v_{gs} + v_{out}$. Therefore, the voltage gain is expressed as

$$A_v = \frac{v_{out}}{v_{in}} = \frac{g_m R_L'}{1 + g_m R_L'}$$

The input resistance is

$$R_{in} = \frac{v_{in}}{i_{in}} = \frac{R_1 R_2}{R_1 + R_2}$$

We follow the same manner as we did in the previous case to calculate the output resistance: zero the independent source and keep the controlled source, then find out the equivalent resistance between the output terminals. It should be noted that $v_{gs} \neq 0$; therefore, to determine the output resistance, we may apply a test voltage source v_x across the output terminals, as shown in Fig. 9.17. Assume the output current of the voltage source is i_x, and the output resistance can be calculated from

$$R_{out} = \frac{v_x}{i_x}$$

Solving for the circuit shown in Fig. 9.17, we have

$$\begin{cases} v_x = \dfrac{i_x + g_m v_{gs}}{1/r_d + 1/R_S} \\ v_x = -v_{gs} \end{cases}$$

Consequently, the output resistance is given by

$$R_{out} = \frac{1}{1/r_d + 1/R_S + g_m}$$

9.4 MOSFET Logic Gates

A most important application of MOSFETs is in logic circuits. In this section, we introduce how to build basic logic gates with MOSFETs.

9.4.1 Inverter

The circuit diagram of an inverter is shown in Fig. 9.18(a). It involves a PMOS and an NMOS, and the DC supply voltage V_{DD} is applied at the drain terminal of the PMOS. When the input voltage is high ($V_{in} = V_{DD}$), PMOS operates in the cutoff region and behaves as an open switch; a conducting channel is induced between the drain and the source of the NMOS, thus it performs as a low resistance and can be modeled by a closed switch. In this case, Fig. 9.18(a) can be replaced by its equivalent circuit, as shown in Fig. 9.18(b), and the output voltage of the circuit is $V_{out} = 0$. When V_{in} is low, a conducting channel is induced under the gate of the PMOS, while the NMOS works in the cutoff region. Hence, the PMOS is on, whereas the NMOS is off. The equivalent circuit is illustrated in Fig. 9.18(c), where the output voltage $V_{out} = V_{DD}$.

Because of the switching action of the transistors, the output voltage is low when the input is high, and vice versa. This is exactly how a logic inverter is supposed to behave.

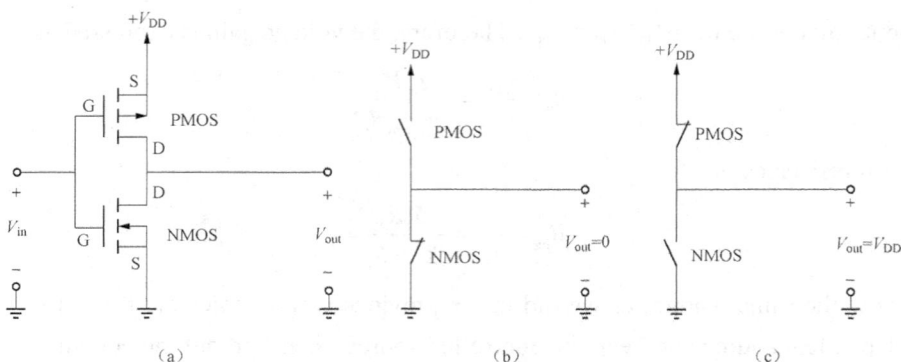

图 9.18 非门

（a）非门的电路图；（b）当 V_{in} 较大时的等效电路；（c）当 V_{in} 较小时的等效电路

Fig. 9.18 The inverter

(a) Circuit diagram of an inverter; (b) Equivalent circuit when V_{in} is high; (c) Equivalent circuit when V_{in} is low

9.4.2 与非门

在非门电路的基础上再加上一对场效应晶体管，即可形成两输入的与非门电路，如图 9.19（a）所示。其中两个 PMOS 管并联，而两个 NMOS 管串联。当输入电压 V_{inA} 和 V_{inB} 均为高时，两个 NMOS 管导通，而两个 PMOS 管断开，等效电路如图 9.19（b）所示，输出电压 $V_{out}=0$。反之，当 V_{inA} 和 V_{inB} 均为低时，两个 NMOS 管断开，而两个 PMOS 管导通，等效电路如图 9.19（c）所示，输出电压 $V_{out} = V_{DD}$。

当输入电压 V_{inA} 高而 V_{inB} 低时，等效电路如图 9.19（d）所示，输出电压 $V_{out} = V_{DD}$。同理，当输入电压 V_{inA} 低而 V_{inB} 高时，也有输出电压 $V_{out} = V_{DD}$。请读者自行画出这种情况下的等效电路。

通过以上分析可见，仅当两个输入电压均为高时，输出电压为零。显然该电路实现了与非门的功能。增加场效应管的数目，可以相应地构造三输入、四输入的与非门。

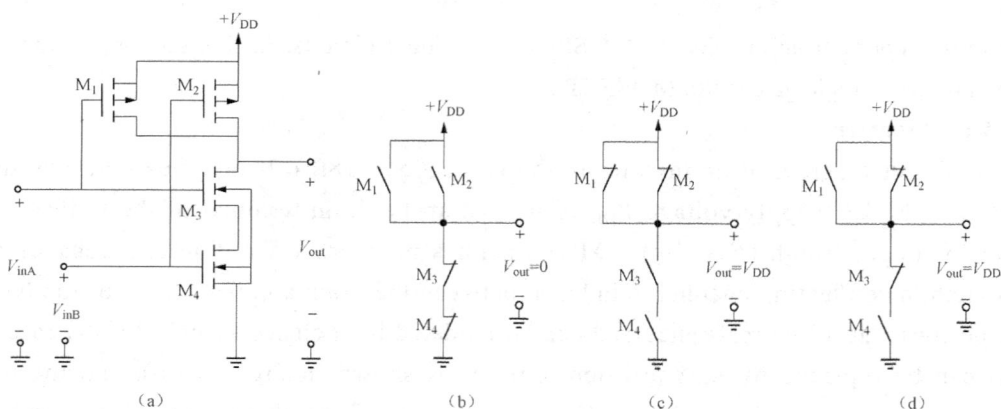

图 9.19 与非门

（a）与非门的电路图；（b）当 V_{inA} 和 V_{inB} 均为高时的等效电路；

（c）当 V_{inA} 和 V_{inB} 均为低时的等效电路；（d）当 V_{inA} 高而 V_{inB} 低时的等效电路

Fig. 9.19 An NAND gate

(a) Circuit diagram of an NAND gate; (b) Equivalent circuit when both V_{inA} and V_{inB} are high;

(c) Equivalent circuit when both V_{inA} and V_{inB} are low; (d) Equivalent circuit when V_{inA} is high while V_{inB} is low

9.4.2   NAND gate

By adding one more pair of PMOS and NMOS transistors to the inverter circuit, we can build a two-input NAND gate, as shown in Fig. 9.19(a). In this circuit, two PMOS transistors are connected in parallel and two NMOS transistors are in series. When the input voltages V_{inA} and V_{inB} are high, both NMOS transistors are on and both PMOS transistors are off. The equivalent circuit is shown in Fig. 9.19(b), where we have the output voltage $V_{out} = 0$. On the contrary, when both input voltages are low, the NMOS transistors are off while the PMOS transistors are on. The equivalent circuit is shown in Fig. 9.19(c), and the output voltage $V_{out} = V_{DD}$.

The equivalent circuit with V_{inA} high and V_{inB} low is given in Fig. 9.19(d), and the output voltage $V_{out} = V_{DD}$. Similarly, when V_{inA} is low and V_{inB} is high, we also have $V_{out} = V_{DD}$. The readers are invited to draw the equivalent circuit under this condition.

From the above analysis, it can be found that the output voltage is zero only when both input voltages are high. Apparently this is exactly what we expect from an NAND gate. By adding more transistors, we could build three-input, four-input NAND gates.

9.4.3   NOR Gate

Building a two-input NOR gate is quite similar to building an NAND gate. As shown in Fig. 9.20(a), the two PMOS transistors are connected in series while the NMOS transistors are in parallel. The analysis of their operation is the same as discussed previously. The equivalent circuits under the conditions of both input voltages being high, both being low, and V_{inA} high while V_{inB} low are shown in Fig. 9.20(b)~(d), respectively. The output voltage under these conditions is left for the readers to analyze.

▶ Problems

P9.1   Determine the region of operation for each of the MOSFETs shown in Fig. P9.1. Suppose $|V_T|$=2 V.

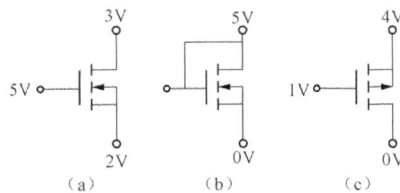

Fig. P9.1

P9.2   For a common-source amplifier shown in Fig. P9.2, where $KP = 50\ \mu A/V^2$, V_T=2 V, $L = 10\ \mu m$, $W = 200\ \mu m$, $r_d = \infty$. Determine whether the NMOS works in the saturation region. If it does, find out its voltage gain, input resistance, and output resistance.

P9.3   Fig. P9.2 shows a common-gate amplifier. Suppose the NMOS has $KP = 50\ \mu A/V^2$, V_T=1 V, $L = 10\ \mu m$, $W = 600\ \mu m$, and $r_d = \infty$. Use large signal analysis to determine the Q point and the voltage of g_m. Use small signal analysis to determine the input resistance and voltage gain.

9.4.3 或非门

或非门电路的工作原理与与非门非常相似，构造双输入或非门电路只需将两个 PMOS 管串联而将两个 NMOS 管并联即可，如图 9.20（a）所示。对于 4 个场效应晶体管的导通/断开状态的分析与前述一致。在输入电压均为高、均为低及 V_{inA} 高、V_{inB} 低这三种情况下的等效电路分别如图 9.20（b）～（d）所示。对于输出电压的分析请读者自行完成。

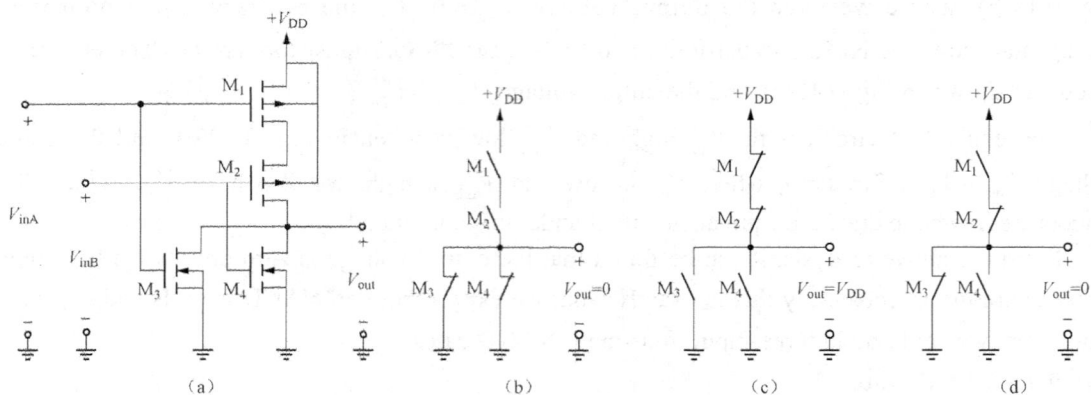

图 9.20 或非门

（a）或非门电路图；（b）当 V_{inA} 和 V_{inB} 均为高时的等效电路；

（c）当 V_{inA} 和 V_{inB} 均为低时的等效电路；（d）当 V_{inA} 高而 V_{inB} 低时的等效电路

Fig. 9.20 An NOR gate

(a) Circuit diagram of an NOR gate; (b) Equivalent circuit when both V_{inA} and V_{inB} are high;

(c) Equivalent circuit when both V_{inA} and V_{inB} are low; (d) Equivalent circuit when V_{inA} is high while V_{inB} is low

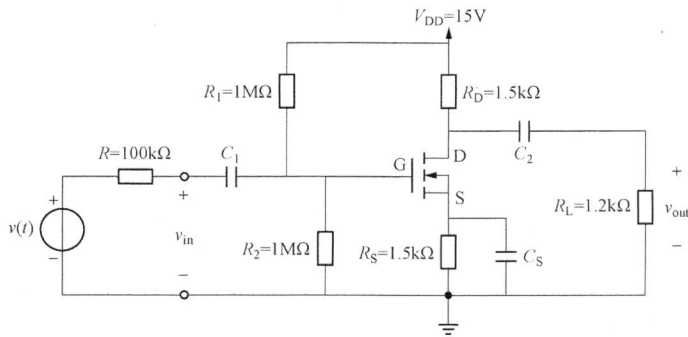

Fig. P9.2

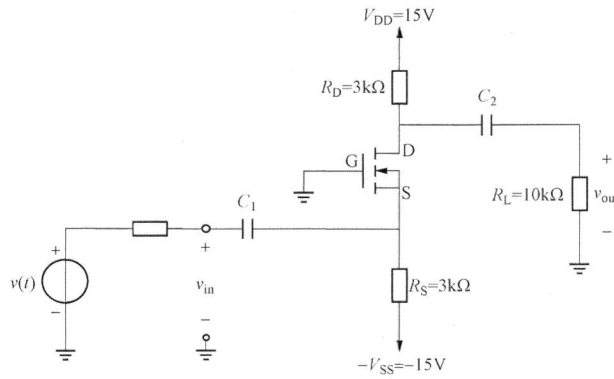

Fig. P9.3

运算放大器
Operational Amplifier

本章将介绍在工程仪表中广泛使用的一个重要器件，运算放大器。运算放大器由大约30个双极型三极管或场效应管、10个电阻和一些电容组成。将这些组成元件通过一定的工序集成在一块晶体上（称为芯片），制造出来的电路称为集成电路。

当前，运算放大器（简称运放）是应用广泛的集成电路。不过，这种类型的放大器最初用于在模拟计算机电路中执行信号的积分或加减法，因此被称为运算放大器。

我们将会看到，这些廉价的集成电路可以与电阻（有些时候是电容）组成很多有用的电路。而且，这些电路的特性主要取决于电路结构和电阻值，而受运放本身一些参数值不稳定的影响很小。

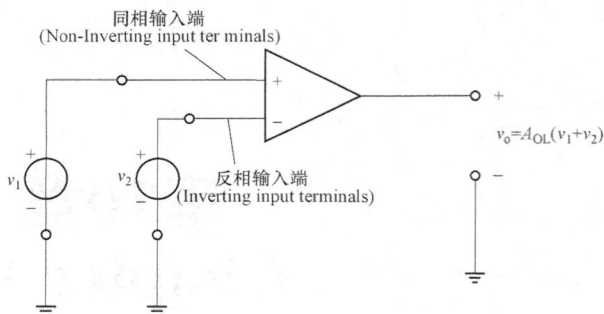

图 10.1　运算放大器的电路符号

Fig. 10.1　Circuit symbol for op amp

10.1　理想运算放大器

运算放大器的电路符号如图 10.1 所示。运算放大器是一个具有反相和同相输入端的差分放大器。输入信号由 $v_1(t)$ 和 $v_2(t)$ 表示。（通常，我们使用小写字母表示随时间变化的电压，可省略时间变量而用 v_1、v_2 等表示电压）。

我们知道输入信号的平均值叫作共模信号，表示为

$$v_{icm} = \frac{1}{2}(v_1 + v_2)$$

同样地，输入电压之间的差叫作差模信号，表示为

$$v_{id} = v_1 - v_2$$

理想运算放大器有以下特性：

（1）输入阻抗为无限大；

（2）差模输入信号增益为无限大；

（3）共模输入信号增益为零；

（4）输出阻抗为零；

（5）无限带宽。

理想运算放大器的等效电路只包含一个受控源，如图 10.2 所示。开环增益 A_{OL} 非常大，理想情况下为无限大。

我们将看到，运算放大器中通常有一部分输出信号引回作为输入信号的反馈网络。因此，信号通过放大器输出，再由反馈网络引回作为输入信号形成一个闭环过程。A_{OL} 是放大器没有反馈网络时的增益，因此称为开环增益。

现在，假设放大器的开环增益 A_{OL} 恒定不变。因此，理想运算放大器没有线性或者非线性的失真，其输出电压 v_o 和差模输入电压 $v_{id}=(v_1-v_2)$ 的波形相同。

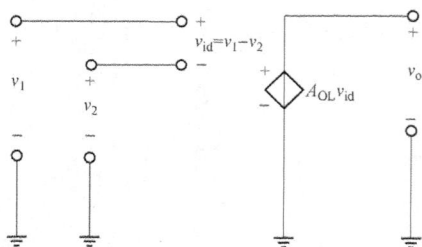

图 10.2　理想运算放大器的等效电路（开环增益 A_{OL} 非常大，接近无限大）

Fig. 10.2　Equivalent circuit for the ideal op amp. The open-loop gain A_{OL} is very large (approaching infinity)

In this chapter, we introduce an important device known as the operational amplifier, which finds application in a wide range of engineering instrumentation. An operational amplifier is a circuit composed of perhaps 30 BJTs or FETs, 10 resistors, and several capacitors. These components are manufactured concurrently on a single piece of silicon crystal (called a chip) by a sequence of processing steps. Circuits manufactured in this way are called integrated circuits (ICs).

Currently, the term operational amplifier, or less formally op amp, refers to the integrated circuits that are employed in a wide variety of general-purpose applications. However, this type of amplifier originated in analog-computer circuits in which it was used to perform such operations as integration or addition of signals; hence, the name operational amplifier.

We will see that inexpensive integrated-circuit op amps can be combined with resistors (and sometimes capacitors) to form many useful circuits. Furthermore, the characteristics of these circuits can be made to depend on the circuits configuration and the resistor values but only weakly on the op amp, which can have large unit-to-unit variations in some of its parameters.

10.1   Ideal Operational Amplifier

The circuit symbol for the operational amplifier is shown in Fig. 10.1. The operational amplifier is a differential having both inverting and non-inverting input terminals. The input signals are denoted as $v_1(t)$ and $v_2(t)$. (As usual, we use lowercase letters to represent general time-varying voltages. Often, we will omit the time dependence and refer to the voltages simply as v_1, v_2, and so on.)

Recall that the average of the input voltages is called the common-mode signal and is given by

$$v_{icm} = \frac{1}{2}(v_1 + v_2)$$

Also, the difference between the input voltages is called the differential signal, given by

$$v_{id} = v_1 - v_2$$

An ideal operational amplifier has the following characteristics.

(1) Infinite input impedance.

(2) Infinite gain for the differential input signal.

(3) Zero gain for the common-mode input signal.

(4) Zero output impedance.

(5) Infinite bandwidth.

An equivalent circuit for the ideal operational amplifier consists simply of a controlled source as shown in Fig. 10.2. The open-loop gain A_{OL} is very large in magnitude-ideally, infinite.

As we will shortly see, op amps are generally used with feedback networks that return part of the output signal to the input. Thus, a loop is created in which signals flow through the amplifier to the output and back through the feedback network to the input. A_{OL} is the gain of the op amp without a feedback network. That is why we call it the open-loop gain.

For now, we assume that the open-loop gain A_{OL} is constant. Thus, there is no distortion, either linear or nonlinear, and the output voltage v_O has a wave-shape identical to that of the differential input $v_{id}=(v_1-v_2)$.

10.2 反相放大器

运算放大器几乎都采用了负反馈，即将输出信号返回到系统的输入端和信号源一起作为输入信号，反馈回来的信号和源信号极性相反（同样存在正反馈，即将输出信号返回到系统的输入信号和源信号一起作为输入信号，反馈回来的信号和源信号极性相同，然而负反馈在放大器电路中应用更广）。通常，会假设运算放大器为理想运放，并使用结合点约束的方法来分析放大器电路。

对于理想运算放大器，其开环差模增益是接近无限大的，甚至一个非常小的输入电压都会得到非常大的输出电压。在一个负反馈电路中，输出信号的一小部分是连接回到反相输入端作为反相输入信号的，这就迫使差模电压趋向于零。如果假设增益是无穷大的，那么差模电压为零，于是，输入电流也为零，这种电压和电流均为 0 的情况称为结合点约束。

理想运算放大器的电路分析步骤如下：

（1）确定有无负反馈；

（2）假设运算放大器的差模电压和输入电流趋于零（结合点约束）；

（3）应用电路分析原理，例如基尔霍夫定律和欧姆定律，求解电路中的各参数。

接下来，我们将展示如何用这种分析方法来分析普遍应用于工程仪表中的重要电路。

10.2.1 基本反相器

运算放大器构成的电路基本反相器如图 10.3 所示。我们通过理想放大器的模型和运行结合点约束来确定电压增益 $A_v = v_o/v_{in}$。但是，在分析电路之前，应首先确认电路引入了负反馈而不是正反馈。

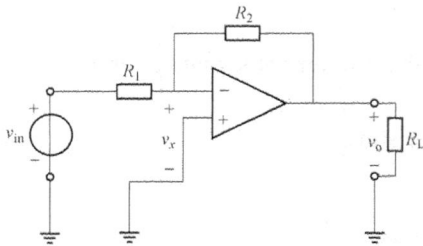

图 10.3 反相放大器

Fig. 10.3 The inverting amplifier

在图 10.3 中，反馈为负反馈。例如，假设某一时刻输入电压 v_{in} 在反相输入端产生了一个正极性电压 v_x。这时，在输出端出现一个非常大（理论上为无穷大）的反相输出电压。而这个输出电压的一部分又通过电阻 R_2 反向分压到反相输入端。因此，反向输入中最初的正极性电压由于负反馈的作用逐渐趋近于零。同理，反相输入端上的初始负极性电压会因引入负反馈而趋于零。因为假设了运算放大器的增益无限大，所以只要有一个非常小（理论上为零）的输入电压 v_x 就可以得到需要的输出了。

图 10.4 给出了一个反相放大器，图中包含了运算放大器输入的结合点约束条件。注意输入电压 v_{in} 加在电阻 R_1 上，因此流过电阻 R_1 的电流为

$$i_1 = \frac{v_{in}}{R_1} \tag{10.1}$$

由于流入运算放大器输入端的电流为零，所以流过电阻 R_2 的电流为

$$i_2 = i_1 \tag{10.2}$$

10.2   Inverting Amplifiers

Operational amplifiers are almost always used with negative feedback, in which part of the output signal is returned to the input in opposition to the source signal. (It is also possible to have positive feedback, in which the signal returned to the input aids the original source signal. However, negative feedback turns out to be more useful in amplifier circuits.) Frequently, we analyze op-amp circuits by assuming an ideal op amp and employing a concept that we call the summing-point constraint.

For an ideal op amp, the open-loop differential gain is assumed to approach infinity, and even a very tiny input voltage results in a very large output voltage. In a negative-feedback circuit, a fraction of the outputs is returned to the inverting input terminal. This forces the differential input voltage toward zero. If we assume infinite gain, the differential input voltage is driven to zero exactly. Since the differential input voltage of the op amp is zero, the input current is also zero. The fact that the differential input voltage and the input voltage and the input current are forced to zero is called the summing-point constraint.

Ideal op-amp circuits are analyzed by the following steps:

(1) Verify that negative feedback is present.

(2) Assume that the differential input voltage and the input current of the op amp are forced to zero (This is the summing-point constraint).

(3) Apply standard circuit-analysis principles, such as Kirchhoff's laws and Ohm's law, to solve for the quantities of interest.

Next, we illustrate this type of analysis for some important circuits that are commonly used in engineering instrumentation.

10.2.1   The basic inverter

An op-amp circuit known as the inverting amplifier is shown in Fig. 10.3. We will determine the voltage gain $A_v = v_o / v_{in}$ by assuming an ideal op amp and employing the summing-point constraint. However, before starting analysis of an op-amp circuit, we should always check to make sure that negative feedback is present rather than positive feedback.

In Fig. 10.3, the feedback is negative, as we shall demonstrate. For example, suppose that due to the input source v_{in}, a positive voltage v_x appears at the inverting input. Then a negative output voltage of large (theoretically infinite) magnitude results at the output. Part of this output voltage is returned to the inverting input by the feedback path through R_2. Thus, the initially positive voltage at the inverting input is driven toward zero by the feedback action. A similar chain of events occurs for the appearance of a negative voltage of the op amp takes precisely the value needed to oppose the source and produce (nearly) zero voltage at the op-amp input. Since we assume that the gain of the op amp is infinite, a negligible (theoretically zero) input voltage v_x is needed to produce the required output.

Fig.10.4 shows the inverting amplifier, including the conditions of the summing-point constraint at the input of the op amp. Notice that the input voltage v_{in} appears across R_1. Thus, the current through R_1 is

$$i_1 = \frac{v_{in}}{R_1} \tag{10.1}$$

Because the current flowing into the op-amp input terminals is zero, the current flowing through R_2 is

$$i_2 = i_1 \tag{10.2}$$

因此，由式（10.1）和式（10.2）可得

$$i_2 = \frac{v_{in}}{R_1}$$

（10.3）

列出包含输入端、电阻 R_2 和运算放大器在内的回路电压方程，得

$$v_o + R_2 i_2 = 0$$

（10.4）

用式（10.3）代替式（10.4）中的 i_2，得到电压增益

$$A_v = \frac{v_o}{v_{in}} = -\frac{R_2}{R_1}$$

（10.5）

A_v 是带有反馈网络的电压增益，称为闭环电压增益。

假设运算放大器为理想的情况下，闭环电压增益仅由电阻的比值决定。这是一个非常理想的情况。注意电压增益是一个负值，这说明放大器是反相放大的（输出电压和输入电压是反相的）。

反相放大器的输入阻抗为

$$Z_{in} = \frac{v_{in}}{i_1} = R_1$$

（10.6）

因此，我们能够通过选择 R_1 的数值方便地调节电路的输入阻抗。

整理式（10.5）可得

$$v_o = -\frac{R_2}{R_1} v_{in}$$

（10.7）

可见输出电压与负载 R_L 无关。从而得到一个结论：输出电压可以看作理想电压源（就它与 R_L 的关系而言）。换句话说，反相放大器的输出阻抗为 0。

10.2.2　虚短

有时，图 10.4 中运算放大器输入端的状态称为虚短。这个术语用于描述运算放大器的差模输入电压趋近于零（如同直接接地）时，运算放大器的输入电压也为零。

如果没有意识到运算放大器的输出通过反馈使差模输入电压为零，则这个概念不易理解。（由于没有电流流入运算放大器，这个概念也可以理解成"虚断"。）

10.2.3　反相放大器的分析

运算放大器可以构造多种反相放大器。反相放大器分析时所遵循的步骤与分析基本反相器的步骤相同：先确定是否存在负反馈，再根据结合点的约束条件，应用基本电路原理分析。

【例 10.1】　反相放大器的分析

图 10.5 是一种采用小阻值电阻构成的高增益放大器。假设运算放大器为理想运放，求其电压增益的表达式，并求出输入阻抗和输出阻抗。已知：$R_1 = R_3 = 1\ \mathrm{k\Omega}$、$R_2 = R_4 = 10\ \mathrm{k\Omega}$。参考图 10.4 所示的常规反相放大器，当 $R_1 = 1\ \mathrm{k\Omega}$ 时，如果要使图 10.4 所示电路达到与图 10.5 所示电路相同的增益，求解需要的 R_2 数值。

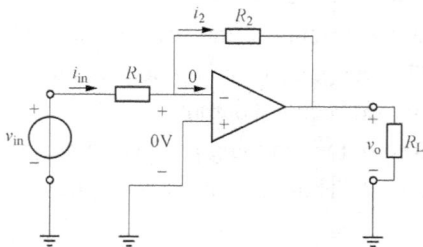

图 10.4　用结合点约束条件分析反相放大器

Fig. 10.4　Using the summing-point constraint in the analysis of the inverting amplifier

Thus, from (10.1) and (10.2),we have

$$i_2 = \frac{v_{in}}{R_1} \qquad (10.3)$$

Writing a voltage equation around the loop by including the output terminals, the resistor R_2, and the op-amp input, we obtain

$$v_o + R_2 i_2 = 0 \qquad (10.4)$$

Using (10.3) to substitute for i_2 in (10.4) and solving for the circuit voltage gain, we have

$$A_v = \frac{v_o}{v_{in}} = \frac{R_2}{R_1} \qquad (10.5)$$

We refer to A_v as the closed-loop gain because it is the gain of the circuit with the feedback network in place.

Under the ideal-op-amp assumption, the closed-loop voltage gain is determined solely by ratio of the resistances. This is a very desirable situation because resistors are available with precise and stable values. Notice that the voltage gain is negative, indicating that the amplifier is inverting (i.e. the output voltage is out of phase with the input voltage).

The input impedance of the inverting amplifier is

$$Z_{in} = \frac{v_{in}}{i_1} = R_1 \qquad (10.6)$$

Thus, we can easily control the input impedance of the circuit by our choice of R_1.

Rearranging (10.5), we have

$$v_o = \frac{R_2}{R_1} v_{in} \qquad (10.7)$$

Consequently, we see that the output voltage is independent of the load resistance R_L. We conclude that the output acts as an ideal voltage source (as far as R_L is concerned). In other words, the output impedance of the inverting amplifier is zero.

10.2.2 Virtual-short-circuit concept

Sometimes, the condition at the op-amp input terminals of Fig. 10.4 is called a virtual **short circuit**. This terminology is used because even though the differential input voltage of the op amp is forced to zero (as if by a short circuit to ground), the op-amp input current is also zero.

This terminology can be confusing unless it is realized that it is the action at the output of the op amp acting through the feedback network that enforces zero differential input voltage. (Possibly, it would be just as valid to call the condition at the op-amp input terminals a "virtual open circuit" because no current flows.)

10.2.3 Variations of the inverter circuit

Several useful versions of the inverter circuit exist. Analysis of these circuits follows the same pattern that we have used for the basic inverter: Verify that negative feedback is present, assume the summing-point constraint, and then apply basic circuit laws.

Example 10.1 Analysis an inverting amplifier

Fig. 10.5 shows a version of the inverting amplifier that can have high gain magnitude without

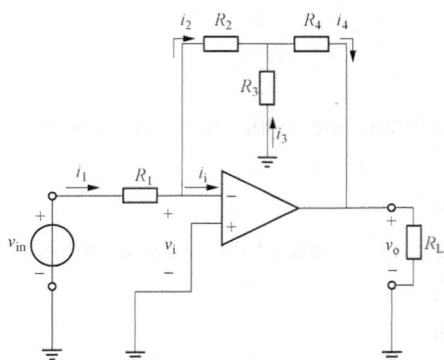

图 10.5 采用小阻值电阻构成的高增益放大器

Fig. 10.5 An inverting amplifier that achieves high gain magnitude with a smaller range of resistance values than required for the basic inverter

解 首先，确定电路中是否存在负反馈。假设输入信号 v_i 为正值，可以得到一个非常大的反相输出电压。其中的一部分通过电阻网络返回输入并和原始的输入信号极性相反。因此，可以证明负反馈的存在。

接下来，我们采用结合点约束条件 $v_i=0$ 且 $i_i=0$，并运用基尔霍夫电流和电压定律以及欧姆定律分析电路。由于 $v_i=0$，v_{in} 是加在 R_1 上的电压，可以得到

$$i_1 = \frac{v_{in}}{R_1} \tag{10.8}$$

然后在 R_1 右边的结合点上应用基尔霍夫电流定律，得到

$$i_2 = i_1 \tag{10.9}$$

这里利用了 $i_i=0$ 的虚断概念。

根据 $v_i=0$，对 R_2 和 R_3 构成的回路可写出如下电压方程：

$$R_2 i_2 = R_3 i_1 \tag{10.10}$$

在 R_3 上方的结合点上利用基尔霍夫电流定律得到

$$i_4 = i_2 + i_3 \tag{10.11}$$

写出 R_L、R_4 和 R_3 所在回路的电压方程

$$v_o = -R_4 i_4 - R_3 i_3 \tag{10.12}$$

接下来，用替代法消去电流变量（i_1，i_2，i_3 和 i_4）得到输出电压和输入电压的关系式。首先从式（10.8）和式（10.9）中得到

$$i_2 = \frac{v_{in}}{R_1} \tag{10.13}$$

之后用式（10.13）代替式（10.10）中的 i_2，整理后得到

$$i_3 = v_{in} \frac{R_2}{R_1 R_3} \tag{10.14}$$

用式（10.13）和式（10.14）消去式（10.11）中的 i_2 和 i_3，得到

$$i_4 = v_{in} \left(\frac{1}{R_1} + \frac{R_2}{R_1 R_3} \right) \tag{10.15}$$

restoring to as wide a range of resistor values as are needed in the standard invert configuration. Derive an expression for the voltage gain under the ideal-op-amp assumption. Also, find the input impedance and output impedance. Evaluate the results for $R_1=R_3=1$ kΩ and $R_2=R_4=10$ kΩ. Then, consider the standard inverter configuration of Fig. 10.4 with $R_1=1$ kΩ, and find the value of R_2 required to achieve the same gain.

Solution:

First, we verify that negative feedback is present. Assume a positive value for v_i, which results in a negative output voltage of very large magnitude. Part of this negative voltage is returned through the resistor network and opposes the original input voltage. Thus, we conclude that negative feedback is present.

Next, we assume the condition of the summing-point constraints

$$v_i=0 \text{ and } i_i=0$$

Then, we apply KCL, KVL, and Ohm's law to analyze the circuit. To begin, we notice that v_{in} appears across R_1 (because $v_i=0$). Hence, we can write

$$i_1 = \frac{v_{in}}{R_1} \tag{10.8}$$

Next, we apply KCL to the node at the right-hand end of R_1. Using the fact,we obtain

$$i_2 = i_1 \tag{10.9}$$

Hence, we have used the fact $i_i=0$.

Writing a voltage equation around the loop through v_i, R_2 and R_3, we obtain

$$R_2 i_2 = R_3 i_1 \tag{10.10}$$

(In writing this equation, we have used the fact that $v_i=0$.) Applying KCL at the top end of R_3 yields

$$i_4 = i_2 + i_3 \tag{10.11}$$

Writing a voltage equation for the loop containing R_L, R_4 and R_3 gives

$$v_o = -R_4 i_4 - R_3 i_3 \tag{10.12}$$

Next, we use substitution to eliminate the current variables (i_1, i_2, i_3 and i_4) and obtain an equation relating the output voltage to the input voltage. First, from (10.8) and (10.9),we obtain

$$i_2 = \frac{v_{in}}{R_1} \tag{10.13}$$

Then, we use (10.13) to substitute for i_2 in (10.10) and rearrange terms to obtain

$$i_3 = v_{in} \frac{R_2}{R_1 R_3} \tag{10.14}$$

Using (10.13) and (10.14) to substitute for i_2 and i_3 in (10.11),we find that

$$i_4 = v_{in} \left(\frac{1}{R_1} + \frac{R_2}{R_1 R_3} \right) \tag{10.15}$$

Finally, using (10.14) and (10.15) to substitute into (10.12), we obtain

$$v_o = -v_{in} \left(\frac{R_2}{R_1} + \frac{R_4}{R_1} + \frac{R_2 R_4}{R_1 R_3} \right) \tag{10.16}$$

Therefore, the voltage gain of the circuit is

$$A_v = \frac{v_o}{v_{in}} = -\left(\frac{R_2}{R_1} + \frac{R_4}{R_1} + \frac{R_2 R_4}{R_1 R_3} \right) \tag{10.17}$$

最后，将式（10.14）和式（10.15）代入式（10.12），得到

$$v_o = -v_{in}\left(\frac{R_2}{R_1} + \frac{R_4}{R_1} + \frac{R_2 R_4}{R_1 R_3}\right) \tag{10.16}$$

因此，电路的电压增益为

$$A_v = \frac{v_o}{v_{in}} = -\left(\frac{R_2}{R_1} + \frac{R_4}{R_1} + \frac{R_2 R_4}{R_1 R_3}\right) \tag{10.17}$$

从式（10.8）中得到输入阻抗

$$R_{in} = \frac{v_{in}}{i_1} = R_1 \tag{10.18}$$

观察式（10.16）可见，输出电压是与负载无关的。因此，对于负载来说，电路的输出是一个理想电压源。换句话说，输出放大器的输出阻抗为零。

按照已知的电阻值（$R_1 = R_3 = 1\ kΩ$ 且 $R_2 = R_4 = 10\ kΩ$）可得

$$A_v = -120$$

在图 10.4 所示的基本反相放大器电路中，电压增益由式（10.5）给出

$$A_v = -\frac{R_2}{R_1}$$

因此，如果 $R_1 = 1\ kΩ$，要得到-120 的电压增益，需要 $R_2 = 120\ kΩ$。为了达到相同的增益，基本反相放大器需要的电阻比为 120∶1，而对于图 10.5 所示的反相电路只需要 10∶1。有时，在电路中保持电阻比尽可能接近 1，会有显著的优点。因此，图 10.5 所示的电路比图 10.4 所示的电路更好。

10.2.4 正反馈

如果在反相放大器中将运算放大器的两个输入端互换，如图 10.6 所示，将会得到正反馈电路，也就是说，反馈信号将起增强原输入信号的作用。这样，如果输入一个正的电压信号，就会得到一个非常大的正输出电压。输出电压的一部分通过反馈网络回到了运放的输入端，因此输入电压会越来越大，输出电压也会越来越大，并且很快达到运算放大器的最大输出值而开始饱和。

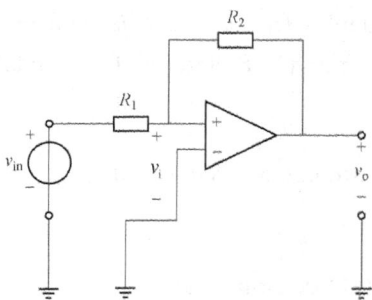

图 10.6　正反馈电路

Fig.10.6　Circuit with positive feedback

如果开始存在一个负的输入电压，那么输出电压就会达到负的最大值。因此，电路就不能起到放大作用——输出电压保持在负最大值或者正最大值，与输入电压 v_{in} 无关。

如果我们忽略图 10.6 所示的电路是正反馈而不是负反馈这个事实，并且错误地运用结合点约束条件，将会得到表达式 $v_o = -(R_2/R_1)v_{in}$，这一点和分析负反馈电路一样。这也说明在使用结合点约束之前验证负反馈是否存在是非常重要的。

The input impedance is obtained from (10.8)

$$R_{in} = \frac{v_{in}}{i_1} = R_1 \qquad\qquad (10.18)$$

Inspection of (10.16) shows that the output voltage is independent of the load resistance. Thus, the output appears as an ideal voltage source to the load. In other words, the output impedance of the amplifier is zero.

Evaluating the voltage gain for the resistor values given ($R_1=R_3=1\text{k}\Omega$ and $R_2=R_4=10\text{ k}\Omega$) yields

$$A_v = 120$$

In the basic inverter circuit of Fig. 10.4, the voltage gain is given by (10.5), which states that

$$A_v = -\frac{R_2}{R_1}$$

Therefore, to achieve a voltage gain of -120 with $R_1=1\text{ k}\Omega$, we need $R_2=120\text{ k}\Omega$. Notice that a resistance ratio of $120:1$ is required for the basic inverter, whereas the circuit of Fig. 10.5 has a ratio of only $10:1$. Sometimes, there are significant practical advantages in keeping the ratio of resistances in circuit as close to unity as possible. Then, the circuit of Fig. 10.5 is preferable to the basic inverter shown in Fig. 10.4.

10.2.4   Positive feedback

It is interesting to consider the inverting amplifier configuration with the input terminals of the op amp interchanged as shown in Fig. 10.6. In this case, the feedback is positive in other words, the feedback signal aids the original input signal. For example, if the input voltage v_i is positive, a very large positive output voltage results. Part of the output voltage is returned to the op-amp input by the feedback network. Thus, the input voltage becomes larger, causing an even larger output voltage. The output quickly becomes saturated at the maximum possible voltage that the op amp can produce.

If an initial negative input voltage is present, the output saturates at its negative extreme. Hence, the circuit does not function as an amplifier-the output voltage is stuck at one extreme or the other and does not respond to the input voltage v_{in}.

If we were to ignore the fact that the circuit of Fig. 10.6 has positive rather than negative feedback and to apply the summing-point constraint erroneously, we could obtain $v_o=-(R_2/R_1)v_{in}$, just as we did for the circuit with negative feedback. This illustrates the importance of verifying that negative feedback is present before using the summing-point constraint.

10.3 同相放大器

同相放大器如图 10.7 所示，假设运算放大器是理想的。首先，确定反馈是正的还是负的。假设 v_i 是正极信号，产生一个非常大的正输出电压。输出电压的一部分通过 R_1 又出现在输入端。因为 $v_i=v_{in}-v_1$，随着 v_o 和 v_1 的增大，v_i 逐渐趋于零。所以，放大器的反馈网络表现为使 v_i 逐渐趋于零。可见，因为反馈信号和原输入信号相反，所以这是负反馈。

接下来，利用结合点约条件 $v_i=0$ 且 $i_i=0$，并利用基尔霍夫电压定律和条件约束 $v_i=0$，得到

$$v_{in} = v_1 \tag{10.19}$$

由于 i_i 为零，R_1 上的电压根据分压公式可得

$$v_1 = \frac{R_1}{R_1 + R_2} v_o \tag{10.20}$$

将式（10.20）代入式（10.19）并整理，得到闭环电压增益为

$$A_v = \frac{v_o}{v_{in}} \tag{10.21}$$

$$A_v = 1 + \frac{R_2}{R_1} \tag{10.22}$$

注意：电路是同相放大器（即 A_v 是正的），同时增益的大小由反馈电阻的比值决定。运算放大器的输入电阻理论上是无穷大的，因为 i_i 是零。既然电压增益不受负载电阻的影响，那么输出阻抗是零。因此，理想运算放大器构成的反相放大器是一个理想的电压放大器。

从式（10.21）中能够看到，当 $R_2=0$ 时增益为 1。但是通常选择断开 R_1 来实现增益为 1，这样的电路称为电压跟随器，如图 10.8 所示。

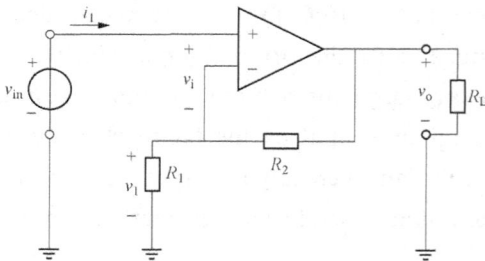

图 10.7　同相放大器

Fig. 10.7　Non-inverting amplifier

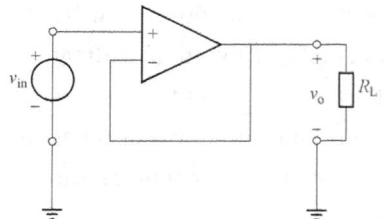

图 10.8　$A_v=1$ 的电压跟随器

Fig. 10.8　The voltage follower which has $A_v=1$

10.4 积分器和微分器

10.4.1 积分器电路

积分器的电路如图 10.9 所示，其输出电压正比于输入电压对运行时间求积分值（对运行时间求积分的意思是指积分的上限为时间 t）。

积分电路在测量仪表中的作用很大。例如，针对一个来自速度传感器的正比于加速度的信号，可以通过对加速度信号进行积分，从而得到一个正比于速度的信号。换一种积分方式可产生正比于位移的信号。

10.3　Non–inverting Amplifiers

The circuit configuration for a non-inverting amplifier is shown in Fig. 10.7. We assume an ideal op amp to analyze the circuit. First, we check to see whether the feedback is negative or positive. In the case, it is negative. To see this, assume that v_i becomes positive and notice that it produces a large positive output voltage. Part of the output voltage appears across R_1. Since $v_i=v_{in}-v_1$, the voltage v_i becomes smaller as v_o and v_1 become larger. Thus, the amplifier and feedback network act to drive v_{in} toward zero. This is negative feedback because the feedback signal opposes the original input.

Having verified that negative feedback is present, we utilize the summing-point constraint: $v_i=0$ and $i_i=0$. Applying KVL and the fact that $v_i=0$, we can write

$$v_{in} = v_1 \tag{10.19}$$

Since i_i is zero, the voltage across R_1 is given by the voltage-division principle

$$v_1 = \frac{R_1}{R_1 + R_2} v_o \tag{10.20}$$

Using (10.20) to substitute into (10.19) and rearranging, we find that the closed-loop voltage gain is

$$A_v = \frac{v_o}{v_{in}} \tag{10.21}$$

$$A_v = 1 + \frac{R_2}{R_1} \tag{10.22}$$

Notice that the circuit is a non-inverting amplifier (A_v is positive), and the gain is set by the ratio of the feedback resistors.

The input impedance of the circuit is theoretically infinite because the input current i_i is zero. Since the voltage gain is independent of the load resistance. Thus, the output impedance is zero. Therefore, under the ideal-op-amp assumption, the non-inverting amplifier is an ideal voltage amplifier.

Notice from (10.21) that minimum gain magnitude is unity, which is obtained with $R_2=0$. Usually, we choose R_1 to be an open circuit for unity gain. The resulting circuit, called a **voltage follower**, is shown in Fig. 10.8.

10.4　Integrators and Differentiators

10.4.1　Integrator circuit

Fig. 10.9 shows the diagram of an **integrator**, which is a circuit that produces an output voltage proportional to the running-time integral of the input voltage. (By the term running time integral, we mean that the upper limit of integration is t.)

The integrator circuit is often useful in instrumentation applications. For example, consider a signal from an accelerometer that is proportional to acceleration. By integrating the acceleration signal, we obtain a signal proportional to velocity. Another integration yields a signal proportional to position.

在图 10.9 中，负反馈通过电容产生。因此，假设运放是理想的，其反相输入端的电压为 0，输入电流为

$$i_{in} = \frac{v_{in}(t)}{R} \tag{10.23}$$

流入理想运放输入端的电流为 0。所以，输入电流 i_{in} 流过电容。假设复位开关在 $t=0$ 时刻打开，于是电容电压在 $t=0$ 时为 0。电容上的电压为

$$v_C(t) = \frac{1}{C} \int_0^t i_{in}(t) \mathrm{d}t \tag{10.24}$$

对于理想放大器输入端到电容再到输入端构成的回路可以写出一个电压方程

$$v_o(t) = -v_C(t) \tag{10.25}$$

将式（10.23）代入式（10.24）中，再将得到的结果代入式(10.25)中，得

$$v_o(t) = -\frac{1}{RC} \int_0^t v_{in}(t) \mathrm{d}t \tag{10.26}$$

因此，输出电压是输入电压对运行时间积分的 $-\frac{1}{RC}$ 倍。如果希望积分器有一个正的增益，可以将积分器级联一个反相放大器。

选择 R 和 C 的值可以调整增益的数值大小。当然选择电容时，为了减少费用、体积和规模，我们希望电容值越小越好。然而，对于一个给定的 1/RC 值，电容越小，电阻就会越大，电流 i_{in} 也就会越小。因此，随着电容的减小，偏置电流的影响会越来越明显。通常在设计中尽量折中。

10.4.2 微分器电路

如图 10.10 所示的是一个微分器电路，其输出电压是输入电压对时间求导数的倍数。用与分析积分器的相似方法对电路进行分析，得到的电路的输出电压为

$$v_o(t) = -RC \frac{\mathrm{d}v_{in}(t)}{\mathrm{d}t} \tag{10.27}$$

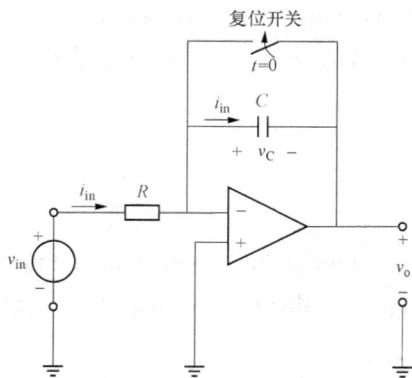

图 10.9　积分器
Fig. 10.9　Integrator

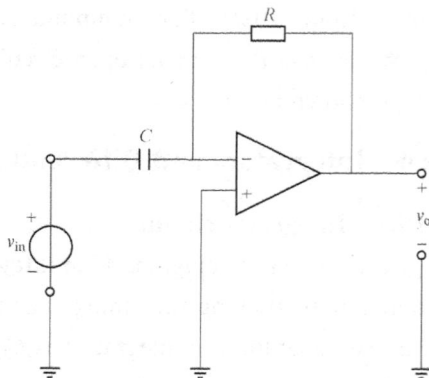

图 10.10　微分器电路
Fig. 10.10　Differentiator

In Fig. 10.9, negative feedback occurs through the capacitor. Thus, assuming an ideal op amp, the voltage at the inverting op-amp input is zero. The input current is given by

$$i_{in} = \frac{v_{in}(t)}{R} \tag{10.23}$$

The current flowing into the input terminal of the (ideal) op amp is zero. Therefore, the input current i_{in} flows through the capacitor. We assume that the reset switch is opened at $t=0$. Therefore, the capacitor voltage is zero at $t=0$. The voltage across the capacitor is given by

$$v_C(t) = \frac{1}{C}\int_0^t i_{in}(t)dt \tag{10.24}$$

Writing a voltage equation from the output terminal through the capacitor and then to ground through the op-amp input terminals, we obtain

$$v_o(t) = -v_C(t) \tag{10.25}$$

Using (10.23) to substitute into (10.24) and the result into (10.25), we obtain

$$v_o(t) = -\frac{1}{RC}\int_0^t v_{in}(t)dt \tag{10.26}$$

Thus, the output voltage is $-1/RC$ times the running integral of the input voltage. If an integrator having positive gain is desired, we can cascade the integrator with an inverting amplifier.

The magnitude of the gain can be adjusted by the choice of R and C. Of course, in selecting a capacitor, we usually want to use as small a value as possible to minimize cost, volume, and mass. However, for a given gain constant $(-\frac{1}{RC})$, smaller C leads to larger R and smaller values of i_{in}. Therefore, the bias current of the op amp becomes more significant as the capacitance becomes smaller. As usual, we try to design for the best compromise.

10.4.2   Differentiator circuit

Fig. 10.10 shows a differentiator that produces an output voltage proportional to the time derivative of the input voltage. By an analysis similar to that used for the integrator, we can show that the circuit produces an output voltage given by

$$v_o(t) = -RC\frac{dv_{in}(t)}{dt} \tag{10.27}$$

▶ Problems

P10.1   A circuit known as a summator is shown in Fig. P10.1. Use the ideal-op-amp assumption to solve for the output voltage in terms of the input voltages and resistor values.

P10.2   Assume an ideal op amp and use the summing-point constraint to find an expression for the output current i_o in the circuit of Fig. P10.2. Also find the input and output resistance of the circuit.

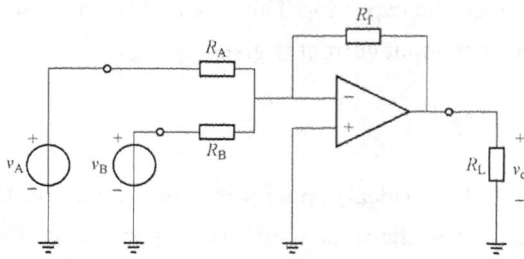

Fig. P10.1

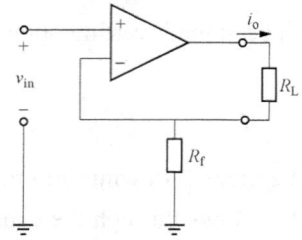

Fig. P10.2

P10.3   Design a circuit with op-amp to implement the following equation:

(1) $v_o = 3(v_1 - v_2) + 5\int_0^t (v_1 - v_2)\mathrm{d}t + 7\dfrac{\mathrm{d}(v_1 - v_2)}{\mathrm{d}t}$

(2) $v_o = 4(v_1 + v_2) - 7\int_0^t (v_1 + v_2)\mathrm{d}t$

逻辑电路
Logic Circuits

到目前为止，前面几章中的信号皆为模拟信号。模拟信号的每一幅值在连续的选值范围内具有独特的意义。例如，位置传感器可以产生一个与位移成比例的模拟信号，信号的每一幅值代表了一个不同的位置。此章将介绍处理数字信号的电路。数字信号只允许若干限定范围的幅值，在限定区间内的每一幅值具有相同的意义。常见的数字信号是只有两个幅度范围的二值信号，相关的信息由逻辑 1 和 0 表示。计算机就是使用数字电路的例子。相比模拟处理方式，数字处理方式有许多重要的优点。

11.1 基本逻辑电路概念

我们在物理系统的设备中经常会遇到模拟信号。例如，在燃烧引擎的气缸中，会有一个压力传感器产生一个与压力呈正比且随时间变化的电压。模拟信号可以被转换成含有相同的虚拟信息的等效数字信号。之后，计算机或者其他数字电路可以处理这些信息。在很多应用之中，可以选择使用数字处理方式或者模拟处理方式。

11.1.1 数字技术的优点

相比模拟信号，数字信号有几个重要的优点。当噪声混入模拟信号之后，几乎不可能确定原始信号中的准确幅值。但是，噪声混入数字信号之后，如果噪声的幅值不太大，仍然能够确定其代表的逻辑数值。图 11.1 描述了这一区别。

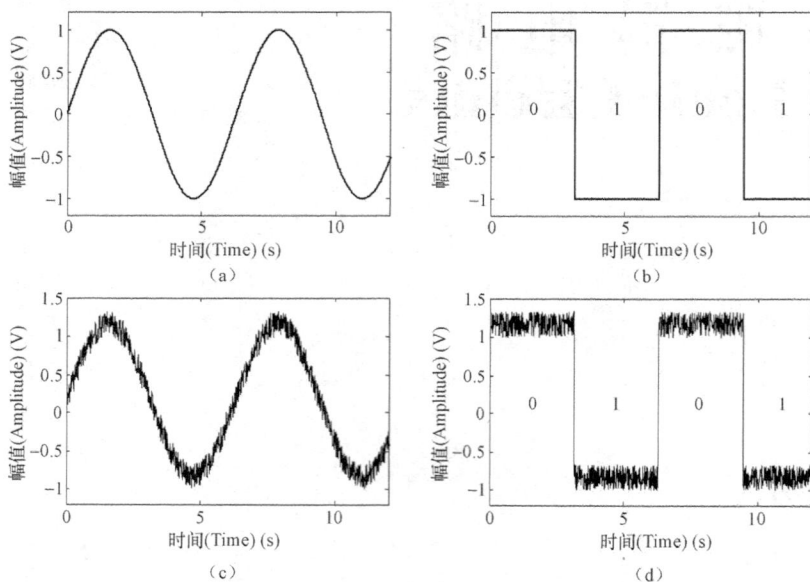

图 11.1　信号与噪声

（a）模拟信号；（b）数字信号；（c）模拟信号加噪声；（d）数字信号加噪声

Fig. 11.1　Signals and noise

(a) An analog signal; (b) A digital signal; (c) An analog signal added noise; (d) A digital signal added noise

对于给定的逻辑电路，较大的电压区间被定义为逻辑 1，其他电压区间被定义为逻辑 0。在正常工作中，逻辑电路仅需产生一个幅值在相应电压区间内的信号就能代表其逻辑意义。因此，在数字电路中的元件电压值不需要像模拟电路中一样精确。

So far, we have considered circuits, such as filters, that process analog signals. For an analogy signal, each amplitude in a continue range has a unique significance. For example, a position sensor may produce an analogy signal that is proportional to displacement. Each amplitude represents a different position. In this chapter, we introduce circuits that process digital signals. For a digital signal, only a few restricted ranges of amplitude are allowed, and each amplitude in a given range has the same significance. Most common are binary signals that take on amplitudes in only two ranges, and the information associated with ranges is represented by the logic values 1 or 0. Computers are examples of digital circuits. We will see that approaches have some important advantages over analog approaches.

11.1　Basic Logic Circuit Concepts

We often encounter analogy signals in instrumentation of physical system. For example, a pressure transducer can yield a voltage that is proportional to pressure versus time in the cylinder of an internal combustion engine. The analog signals can be converted to equivalent digital signals that contain virtually the same information. Then, computers or other digital signals that contain virtually the same information. Then, computers or other digital circuits can be used to process this information. In many applications, we have a choice between digital and analog approaches.

11.1.1　Advantages of the digital approach

Digital signals have several important advantages over analogy signals. After noise is added to an analog signal, it is usually impossible to determine the precise amplitude of the original signal. On the other hand, after noise is added to a digital signal, we can still determine the logic values-provided that the noise amplitude is not too large. This is illustrated in Fig. 11.1.

For a given type of logic circuit, one large of voltages represents logic 1, and another range of voltages represents logic 0. For proper operation, a logic circuit only needs to produce a voltage somewhere in the correct range. Thus, component values in digital circuits do not need to be as precise as in analog circuits.

More facts about ADC

Signals acquired by sensors, such as a sound picked up by a microphone or light entering a digital camera, are analog signals. They can be converted into digital signals through an analog-to-digital converter (ADC). Typically, the digital output is a two's complement binary number that is proportional to the input. The resolution of the ADC refers to the number of discrete values it can produce over the range of analog values. A higher resolution indicates a more precise representation of the analog signal.

在现代集成电路（IC）生产技术中，非常复杂的数字电路（集成了大量的元器件）可以以成本非常低的方式生产。而模拟电路需要大容量的高精度元件才能产生精确的元件电压，导致大规模的模拟电路几乎不可能生产。因此，数字系统在过去的数十年间得到了广泛的发展，未来这一趋势将会一直持续下去。

11.1.2 电平区与噪声区

逻辑电路的输入电压值如果分别在一定允许范围内，则被称为逻辑 1，或与之不重叠的逻辑 0。其中，逻辑 0 或者低电平的允许电压最大值定义为 V_{IL}，逻辑 1 或者高电平的允许电压最小值定义为 V_{IH}，如图 11.2 所示。在 V_{IL} 与 V_{IH} 之间的电压是无意义的，仅会呈现于电压变化的过程中。

此外，电路设计中输出电压的允许范围要比输入电压更窄。如图 11.2 所示，V_{OL} 是低电平输出电压的最大值，V_{OH} 是高电平输出电压的允许最小值。

由于噪声可能在信号传输过程中混入逻辑信号，因此输出电压必须有比输入电压更窄的允许变化范围，这样的差异被称为**噪声容限**。在理想情况下，噪声容限电压越大越好。

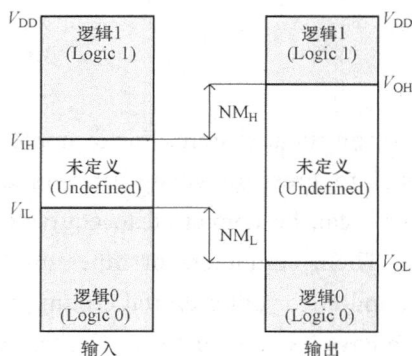

图 11.2 逻辑电路输入与输出的电压范围

Fig. 11.2 Voltage ranges for logic-circuit inputs and outputs

11.2 组合逻辑电路

在本节中，我们将介绍逻辑门电路，即由多个逻辑输入变量产生一个逻辑输出变量的电路。我们将关注于逻辑门的外部特性。下面即将介绍的电路是无记忆的，即它们在瞬时的输出值仅与同一瞬时的输入值有关。

11.2.1 逻辑门

1. 与门

与运算是一个重要的逻辑运算。A 和 B 两个逻辑变量参与的与运算记作 $A \cdot B$，简写为 AB，读作 "A 与 B"。与运算也被称为逻辑乘法。

一种描述组合逻辑系统的方法是将所有的输入变量和对应的输出变量的值列出，这就是真值表。与操作的真值表列于图 11.3（a）中。注意，仅当 A 和 B 同时为 1 时，AB 才为 1；否则，AB 的值为 0。

对于与操作，我们可以得到以下关系式：

$$AA=A \tag{11.1}$$

$$A1=A \tag{11.2}$$

$$A0=0 \tag{11.3}$$

$$AB=BA \tag{11.4}$$

$$A(BC) = (AB)C = ABC \tag{11.5}$$

二输入与门的电路符号如图 11.3（b）所示。

AND 可以有多于两个的输入变量，例如，图 11.4 展示了三输入与门的真值表及其电路符号。

It turns out that with modern integrated-circuit (IC) manufacturing technology, very complex digital logic circuits (containing of components) can be produced economically. Analog circuits often call for large capacitances and precise component values that are impossible to manufacture as large-scale integrated circuits. Thus, digital systems have become increasingly important in the past few decades, a trend that will continue.

11.1.2 Logic ranges and noise margins

Logic circuits are typically designed so that a range of input voltages is accepted as logic 1 and another non-overlapping range of voltages is accepted as logic 0. The largest input voltage accepted as logic 0, or low, is denoted as V_{IL}; and the smallest input voltage accepted as logic 1, or high, is denoted as V_{IH}. This is illustrated in Fig. 11.2. No meaning is assigned to voltages between V_{IL} and V_{IH}, which normally occurs only during transitions.

Furthermore, the circuits are designed so that the output voltages fall into narrower ranges than the inputs (provided that the inputs are in the acceptable ranges). This is also illustrated in Fig. 11.2. V_{OL} is the highest logic-0 output voltage, and V_{OH} is the lowest logic-1 output voltage.

Because noise can be added to a logic signal in the interconnections between outputs and inputs, it is important that the outputs have narrower ranges than the acceptable inputs. The differences are called **noise margins**. Ideally, noise margins are required to be as large as possible.

11.2 Combinatorial Logic Circuits

In this section, we consider circuits called **logic gates** that combine several logic-variable inputs to produce a logic-variable output. We focus on the external behavior of logic gates. The circuits that we are about to discuss are said to be memory-less because their output values at a given instant depend only on the input values at that instant.

11.2.1 Logic gates
1. AND gate

An important logic function is called the AND operation. The AND operation on two logic variables, A and B, is represented as AB, read as "A and B". The AND operation is also called **logical multiplication**.

One way to specify a combinatorial logic system is to list all the possible combinations of the input variables and the corresponding output values. Such a listing is called a **truth table**. The truth table for the AND operation of two variables is shown in Fig. 11.3 (a). Notice that AB is 1 if and only if A and B are both 1; otherwise, AB is 0.

For the AND operation, we can write the following relations:

$$AA = A \tag{11.1}$$
$$A1 = A \tag{11.2}$$
$$A0 = 0 \tag{11.3}$$
$$AB = BA \tag{11.4}$$
$$A(BC) = (AB)C = ABC \tag{11.5}$$

The circuit symbol for a two-input AND gate is shown in Fig. 11.3(b). It is possible to have AND gates with more than two inputs. For example, the truth table and circuit symbol for a three-input AND gate are shown in Fig. 11.4, respectively.

2. 非门

在逻辑变量的符号上方加一条横线便是对该变量取非运算。如果符号 $\overline{A}$ 表示对变量 A 取反，读作 "A 的非" 或 "A 的反"。如果 A 为 0，则 $\overline{A}$ 为 1；反之亦然。

A	B	$C=AB$
0	0	0
0	1	0
1	0	0
1	1	1

（a）　　　　　　　（b）

图 11.3　二输入与门

（a）真值表；（b）符号

Fig. 11.3　Two in puts AND gates

(a) Truth table; (b) Symbol

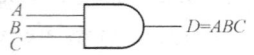

A	B	C	$D=ABC$
0	0	0	0
0	0	1	0
0	1	0	0
0	1	1	0
1	0	0	0
1	0	1	0
1	1	0	0
1	1	1	1

（a）　　　　　　　（b）

图 11.4　三输入与门

（a）真值表；（b）符号

Fig. 10.4　Three inputs AND gates

(a) truth table; (b) Symbol

对变量取反运算的电路称为非门，非门的真值表及符号如图 11.5 所示。符号输出端的小圆圈表示取反。

对于取反运算有以下表达式

$$A\overline{A} = 0 \tag{11.6}$$

$$\overline{\overline{A}} = A \tag{11.7}$$

3. 或门

逻辑变量的或运算写作 $A+B$，读作 "A 或 B"。图 11-6 给出了二输入或门的真值表和电路符号。当 A 或 B 任一为 1 或者两者同时为 1 时，$A+B$ 的值为 1；当两者同时为 0 时，$A+B$ 的值才为 0。或操作也称为逻辑加法。

A	$\overline{A}$
0	1
1	0

（a）　　　　　　　（b）

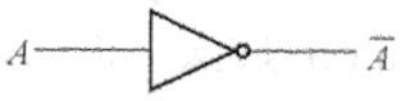

图 11.5　非门

（a）真值表；（b）符号

Fig. 11.5　Logic inverter

(a) Truth table; (b) Symbol

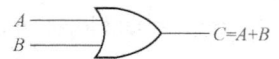

A	B	$C=A+B$
0	0	0
0	1	1
1	0	1
1	1	1

（a）　　　　　　　（b）

图 11.6　二输入或门

（a）真值表；（b）符号

Fig. 11.6　Two inputs OR gates

(a) Truth table; (b) Symbol

对于或操作，有以下关系式：

$$A + 0 = A \tag{11.8}$$

$$A + 1 = 1 \tag{11.9}$$

$$A + \overline{A} = 1 \tag{11.10}$$

$$A + A = A \tag{11.11}$$

$$(A+B)+C = A+(B+C) = A+B+C \tag{11.12}$$

$$A(B+C) = AB + AC \tag{11.13}$$

2. Logic inverter

The NOT operation on a logic variable is represented by placing a bar over the symbol for the logic variable. The symbol $\overline{A}$ is read as "not A " or as " A inverse". If A is 0, $\overline{A}$ is 1; and vice versa.

Circuits that perform the NOT operation are called inverters. The truth table and circuit symbol for an inverter are shown in Fig. 11.5 (Logic inverters). The bubble placed at the output of the inverter symbol is used to indicate inversion.

We can readily establish the following operations for the NOT operation

$$A\overline{A} = 0 \tag{11.6}$$

$$\overline{\overline{A}} = A \tag{11.7}$$

3. OR gate

The OR operation of logic variables is written as $A+B$, which is read as "A or B". The truth table and the circuit symbol for a two-input OR gate are shown in Fig. 11.6. Notice that $A+B$ is 1, if either A or B is 1, or both A and B are 1; $A+B$ is 0 if both A and B are 0. The OR operation is also called logical addition.

For the OR operation, we can write the following relations

$$A + 0 = A \tag{11.8}$$

$$A + 1 = 1 \tag{11.9}$$

$$A + \overline{A} = 1 \tag{11.10}$$

$$A + A = A \tag{11.11}$$

$$(A + B) + C = A + (B + C) = A + B + C \tag{11.12}$$

$$A(B + C) = AB + AC \tag{11.13}$$

George Boole (2 November 1815 – 8 December 1864) was an English mathematician, philosopher and logician. He worked in the fields of differential equations and algebraic logic, and helped establish modern symbolic logic. His algebra of logic, now called Boolean algebra, is basic to the design of digital computer circuits. Boole is now best known as the author of 'An Investigation of the Laws of Thought on Which are Founded the Mathematical Theories of Logic and Probabilities'.

11.2.2 布尔代数

逻辑变量的数学理论称为布尔代数，它是由数学家乔治布尔提出的。为了证明布尔代数中的关系式成立，可以列出包含各变量所有可能的组合真值表，再比较关系式两边是否相等。

1. 布尔代数式的实现

布尔代数式可通过与门、或门和非门的相互连接来实现。例如，逻辑表达式

$$F = A\bar{B}C + ABC + (C+D)(\bar{D}+E) \tag{11.14}$$

可以由图 11.7 所示的电路图实现。

$$F = A\bar{B}C + ABC + (C+D)(\bar{D}+E)$$

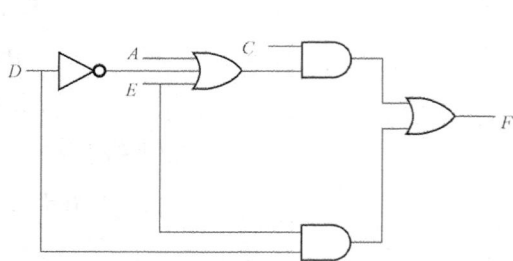

通常可以对逻辑表达式进行化简。例如，在式（11.14）的右侧的最后一项可以展开得到

$$F = A\bar{B}C + ABC + C\bar{D} + CE + D\bar{D} + DE \tag{11.15}$$

由于 $D\bar{D}$ 的逻辑值始终为 0，因此可以从表达式中去掉。再将式（11.15）的右边前两项合并，得到

$$F = AC(\bar{B}+B) + C\bar{D} + CE + DE \tag{11.16}$$

又由于 $\bar{B}+B$ 始终为 1，所以我们能够得到

$$F = AC + C\bar{D} + CE + DE \tag{11.17}$$

再将 C 从前三项中提出，我们能够得到

$$F = C(A + \bar{D} + E) + DE \tag{11.18}$$

该式可以由图 11.8 所示的逻辑电路实现。

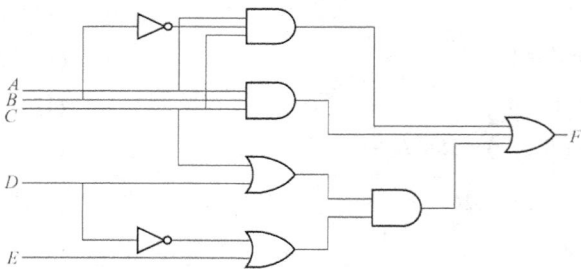

图 11.7　实现逻辑表达式
$F = A\bar{B}C + ABC + (C+D)(\bar{D}+E)$ 的电路
Fig. 11.7　The circuit for the logic equation
$F = A\bar{B}C + ABC + (C+D)(\bar{D}+E)$

图 11.8　由图 11.7 化简得到的电路
Fig. 11.8　The simplified circuit for Fig. 11.7

通常实现给定逻辑函数的逻辑电路有多种。下面将介绍用最少的给定类型的门电路实现逻辑函数的方法。

2. 摩根法则

摩根法则是布尔代数中的两个重要结论，可以表示为

$$ABC = \overline{\bar{A} + \bar{B} + \bar{C}} \tag{11.19}$$

和

$$A + B + C = \overline{\bar{A}\bar{B}\bar{C}} \tag{11.20}$$

摩根法则的另一种表述为：要使逻辑表达式中的每个变量用它们的"非"代替，可将式中的与运算与或运算互换，并将整个表达式取反，这样的逻辑表达式与原表达式等值。

11.2.2   Boolean algebra

The mathematical theory of logic variables is called **Boolean algebra**, named for mathematician George Boole. One way to prove a Boolean algebra identity is to produce a truth table that lists all possible combinations of the variables and to show that both sides of the expression yield the same results.

1.   Implantation of Boolean expressions

Boolean algebra expressions can be implemented by interconnection of AND gates, OR gates, and inverters. For example, the logic expression

$$F = A\bar{B}C + ABC + (C + D)(\bar{D} + E) \tag{11.14}$$

can be implemented by the logic circuit shown in Fig. 11.7.

Sometimes, we can manipulate a logic expression to find an equivalent expression that is simpler. For example, the last term on the right-hand side of equation (11.14) can be expanded, resulting in

$$F = A\bar{B}C + ABC + C\bar{D} + CE + D\bar{D} + DE \tag{11.15}$$

Since the term $D\bar{D}$ always has the logic value 0, it can be dropped from the expression. Factoring the first two terms on the right-hand side of (11.15) results in

$$F = AC(\bar{B} + B) + C\bar{D} + CE + DE \tag{11.16}$$

However, the quantity $\bar{B} + B$ always equals to 1, so we can write

$$F = AC + C\bar{D} + CE + DE \tag{11.17}$$

Factoring C from the first three terms on the right-hand side, we have

$$F = C(A + \bar{D} + E) + DE \tag{11.18}$$

This can be implemented as shown in Fig. 11.8.

Thus, we can often find alternative implementations for a given logic function. Later, we consider methods for finding the implementation using the fewest gates of given type.

2.   De Morgan's laws

Two important results in Boolean algebra are De Morgan's laws, which are given by

$$ABC = \overline{\bar{A} + \bar{B} + \bar{C}} \tag{11.19}$$

and

$$A + B + C = \overline{\bar{A}\bar{B}\bar{C}} \tag{11.20}$$

Augustus De Morgan (27 June, 1806 – 18 March, 1871) is a British mathematician and logician. He formulated De Morgan's laws and introduced the term mathematical induction, making its idea rigorous.

【例 11.1】 应用摩根法则对 $D = AC + \bar{B}C + \bar{A}(\bar{B} + BC)$ 右侧的逻辑表达式进行变换。

解 首先，我们用每个变量的"非"代替这些变量，可得

$$\bar{A}\bar{C} + B\bar{C} + A(B + \bar{B}\bar{C})$$

之后，我们将与运算和或运算互换，可得

$$(\bar{A} + \bar{C})(B + \bar{C})\left[A + B(\bar{B} + \bar{C}) \right]$$

最后，对整个表达式求"非"即可

$$D = \overline{(\bar{A} + \bar{C})(B + \bar{C})[A + B(\bar{B} + \bar{C})]}$$

因此，摩根法则可以让我们用另一方式描述逻辑表达式。

摩根法则表明，任何逻辑函数均可以由与门和非门相互连接实现，式（11.20）说明可以用与操作和非操作替换或操作。类似地，任何逻辑公式可以由或门和非门组成，因为式（11-19）表明可以用或操作和非操作替换与操作。因此，我们仅需要与门或者或门以及非门就能实现任何逻辑表达式，而不必用到全部。

11.2.3　与非门、或非门和异或门

图 11.9 给出了额外的一些逻辑门。与非门等价于一个与门和一个非门串联。注意：在符号上，与非门仅是在与门的输出端上加上了一个小圆圈，以表示在与操作后取了反。类似地，或非门等价于或门和一个非门串联。

$A \oplus B$ 表示 A 和 B 两个逻辑值的异或，其定义为

$$0 \oplus 0 = 0$$
$$1 \oplus 0 = 1$$
$$0 \oplus 1 = 1$$
$$1 \oplus 1 = 0$$

注意：仅当 A 或 B 两个逻辑值有一个为 1 的时候，其异或的结果才为 1。当两者均为 1 的时候，其异或结果为 0。异或运算也称为模 2 加法。

缓冲器只有一个输入端和一个输出端，输出端产生和输入端相同的值。在逻辑电路带有低阻抗负载的时候，常使用缓冲器来提供大电流。

同或门在两个变量相同的时候，输出一个高电平。实际上同或门是由异或门加上一个非门构成，符号如图 11.9（e)所示。

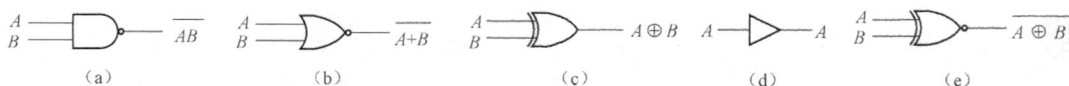

图 11.9　其他的逻辑门符号

（a）与非门；（b）或非门；（c）异或门；（d）缓冲器；（e）同或门

Fig. 11.9　Additional Logic-gate symbols:

(a) NAND gate; (b) NOR gate; (c) XOR gate; (d) Buffer; (e) Equivalance gate (XNOR)

11.2.4　电路的与非门和或非门实现

我们能够看出，通过组合几个不同的门电路可以实现相同的逻辑函数。例如，将与非门的两个输入端连接起来就得到了一个非门，即

$$\overline{(AA)} = \bar{A}$$

Another way to state these laws is as follows: If the variables in a logic expression are replaced by their inverses, the AND operation is replace by OR, the OR operation is replaced by AND, and the entire expression is inverted, the resulting logic expression yields the same values as before the changes.

Example 11.1: Apply De Morgan's laws to the right-hand side of the logic expression

$$D = AC + \bar{B}C + \bar{A}(\bar{B} + BC)$$

Solution:

First, we replace each variable by its inverse, resulting in the expression

$$\bar{A}\bar{C} + B\bar{C} + A(B + \bar{B}C)$$

Then, we replace the AND operation by OR, and vice versa:

$$(\bar{A} + \bar{C})(B + \bar{C})\left[A + B(\bar{B} + \bar{C}) \right]$$

Finally, inverting the expression, we can write

$$D = \overline{(\bar{A} + \bar{C})(B + \bar{C})[A + B(\bar{B} + \bar{C})]}$$

Therefore, De Morgan's laws given us an alternative way to write logic expressions.

An important implication of De Morgan's laws is that we can implement any logic function by using AND gates and inverters. This is true because (11.20) can be employed to replace the OR operation by the AND operation (and logical inversions). Similarly, any logic function can be implemented with OR gates and inverters, because (11.19) can be used to replace AND operation with the OR operation (and logical inversions). Consequently, to implement a logic function, we need inverters and either AND gates or OR gates, rather than both.

11.2.3   NAND, NOR, and XOR gates

Some additional logic gates are shown in Fig. 11.9. The NAND gate is equivalent to an AND gate followed by an inverter. Notice that the symbol is the same as for an AND gate, with a bubble at the output terminal to indicate that the output has been inverted after the AND operation. Similarly, the NOR gate is equivalent to an OR gate followed by an inverter.

The exclusive-OR (XOR) operation for two logic variables A and B is represented by $A \oplus B$ and is defined by

$$0 \oplus 0 = 0$$
$$1 \oplus 0 = 1$$
$$0 \oplus 1 = 1$$
$$1 \oplus 1 = 0$$

Notice that the XOR operation yields 1 if A is 1 or B is 1, but yields 0 if both A and B are 1.The XOR operation is also known as modulo-two addition.

A buffer has a single input and produces an output with the same value as the input. Buffers are commonly used to produce large currents when a logic signal must be applied to a low-impedance load.

The equivalence gate produces a high output only if both inputs have the same value. In effect, it is an XOR followed by an inverter as the symbol of Fig. 11.9(e) (Equivalence gate (XNOR)) implies.

11.2.4   Logical sufficiency of NAND gates and of NOR gates

As we have seen, several combinations of gates often can be found that perform the same function. For example, if the inputs to a NAND are tied together, an inverter results. This is true because

$$\overline{(AA)} = \bar{A}$$

如图 11.10（a）所示。

此外，由摩根法则可知，与运算等效于与非门后面再连上一个与非门，如图 11.10（b）所示；或运算等效于将两个输入变量取反后再加到与非门的两个输入端，如图 11.10（c）所示；与非门后面再连接一个非门可形成一个与门。这样，基本的逻辑函数（与、或和非）均可以由与非门实现。因此可以得出结论：由与非门可以实现任何的组合逻辑函数。

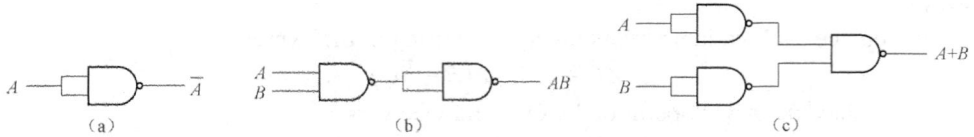

图 11.10　由与非门表示的基本布尔运算

（a）非门；（b）与门；（c）或门

Fig. 11.10　Basic Boolean operations can be implemented with NAND gates：

(a) Inverter; (b) AND gate; (c) OR gate

11.3　逻辑电路的综合

在本节中，将介绍数据输入及指定的具体输出实现逻辑电路的方法。通常情况下，逻辑电路的初始说明是通过自然语言描述的。通过自然语言翻译为真值表或者布尔逻辑表达式，可以求得实际的逻辑电路。

11.3.1　与或式的电路实现

见表 11.1，A、B 和 C 是输入逻辑变量，D 是相应的逻辑输出。注意这里的每一行是按照与 ABC 所表示的二进制对应的十进制数顺序编号的。

表 11.1　　　　　　　用于说明与或式和或与式表达式的真值表

Table 11.1　　　　　Truth table used to illustrate SOP and POS logical expressions

序号（Row）	A	B	C	D
0	0	0	0	1
1	0	0	1	0
2	0	1	0	1
3	0	1	1	0
4	1	0	0	0
5	1	0	1	0
6	1	1	0	1
7	1	1	1	1

为了求得产生输出变量 D 的逻辑电路，需先求出 D 的逻辑表达式。一种方法是通过集合真值表中 D 为 1 的行写出 D 的逻辑表达式。首先在真值表中选出 D 为 1 的行，即第 0、2、6 和 7 行，然后写出这些行的输入逻辑变量的积，并将输入逻辑变量为 0 的取反，使该逻辑积为 1。每一项逻辑积中应包含所有的输入量。例如，逻辑积 $\overline{A}\overline{B}\overline{C}$ 的值仅在第 0 行为 1，逻辑积 ABC 的值仅在第 7 行为 1。包含所有输入变量（或其反）的积项称为**最小项**。

最后，将所有最小项相或就得到输出变量的逻辑表达式，对表 11.1，有

$$D = \overline{A}\overline{B}\overline{C} + \overline{A}B\overline{C} + AB\overline{C} + ABC \tag{11.21}$$

which is illustrated in Fig. 11.10 (a).

Furthermore, as shown by De Morgan's laws, the AND operation can be realized by the cascade connection of two NAND gates, as shown in Fig. 11.10(b). The OR operation can be realized by inverting the input variables and combining the results in a NAND gate. This is shown in Fig. 11.10(c), in which the inverters are formed from NAND gates. Finally, a NAND gate followed by an inverter results in an AND gate. Since the basic logic functions (AND, OR, and NOT) can be realized by using only NAND gates, we conclude that NAND gates are sufficient to realize any combinatorial logic function. Therefore, any Boolean function can be implemented by the use of NAND gates alone.

11.3 Synthesis of Logic Circuits

In this section, we consider methods to implement logic circuits given the specification for the output in terms of the inputs. Often, the initial specification for a logic circuit is given in natural language. This is translated into a truth table or a Boolean logic expression that can be manipulated to find a practical implementation.

11.3.1 Sum-of-products implementation

Consider the truth table shown in Table 11.1, where A, B and C are input logic variables, and D is the desired output. Notice that we have numbered the rows of the truth table with the decimal number corresponding to the binary number formed by ABC.

Suppose that we want to find a logic circuit that produces the output variable D. One way to write a logic expression for D is to concentrate on the rows of the truth table for which D is 1. In Table 11.1, these are the rows numbered 0, 2, 6, and 7. Then, we write a logical product of the input logic variables or their inverses that equals 1 for each of these rows. Each input variable or its inverse is included in each product. In writing the product for each row, we invert the logic variables that are 0 in that row. For example, the logical product $\overline{A}\overline{B}\overline{C}$ equals logic 1 only for row 0, and ABC equals 1 only for row 7. Product terms that include all of the input variables (or their inverses) are called **minterms**.

Finally, we write an expression of the output variable as a logical sum of minterms. For Table 11.1, it yields

$$D = \overline{A}\overline{B}\overline{C} + \overline{A}B\overline{C} + AB\overline{C} + ABC \tag{11.21}$$

Maurice Karnaugh (born in 4 October 4 1924) is an American physicist and mathematician known for the Karnaugh map used in Boolean algebra. Obtaining a Ph.D. degree in physics from Yale University, Karnaugh worked at Bell Laboratories, developing the Karnaugh map as well as patents for Pulse-code modulation (PCM) encoding and magnetic logic circuits and coding.

这种类型的表达式称为与或式。由上述可知，根据给定的真值表，总能找到一个逻辑输出的与或式表达式。由式（11.21）可以得到如图 11.11 所示的逻辑电路图。

与或式的简写法是列出真值表中输出为 1 的行号，写作

$$D = \sum m(0,2,6,7) \tag{11.22}$$

其中 m 表示所列举行对应的最小项。

11.3.2　或与式电路实现

求出输出变量 D 的逻辑表达式的另一种方法是集合真值表中 D 为 0 的行。首先在真值表 11.1 中选出 D 为 0 的行，即第 1、3、4 和 5 行，然后写出这些行的输入逻辑变量的逻辑和，并将输入逻辑变量为 1 的取反，使该逻辑和为 0，每一项逻辑和中应包含所有的输入变量。例如，第 1 行的 $(A+B+\bar{C})$ 逻辑和为 0，第 3 行的 $(A+\bar{B}+\bar{C})$ 逻辑和为 0。包含所有输入变量（或其反）的和项称为**最大项**。

最后，将所有最大项相与就得到输出变量的逻辑表达式。对于表 11.1，有

$$D = (A+B+\bar{C})(A+\bar{B}+\bar{C})(\bar{A}+B+C)(\bar{A}+B+\bar{C}) \tag{11.23}$$

这样的表达式称为或与式。根据给定的真值表，总能找到一个逻辑输出的或与式表达式。图 11.12 给出了式（11.23）所表达的电路。

或与式的简写法是列出真值表中输出为逻辑 0 的行号。这样，式（11.23）可以写为

$$D = \prod M(1,3,4,5) \tag{11.24}$$

式中，M 为所列举行对应的最大项。

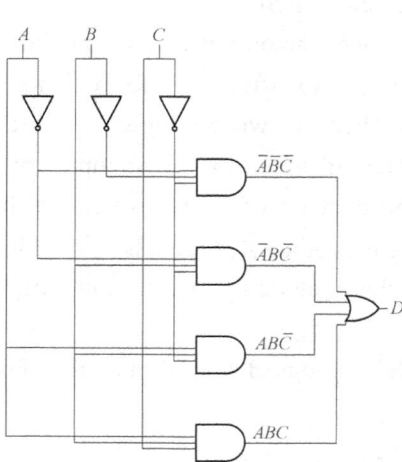

图 11.11　表 11.1 的与或式电路实现

Fig. 11.11　Sum-of-product logic circuit for Table 11.1

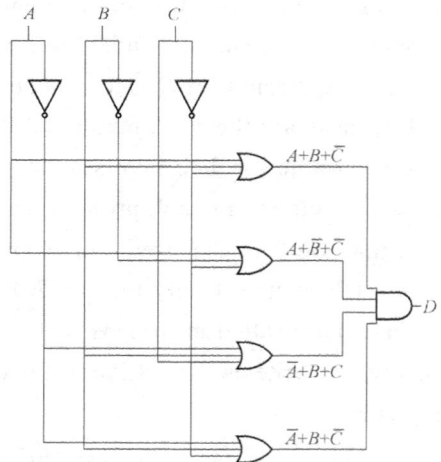

图 11.12　表 11.1 的或与式电路实现

Fig. 11.12　Product-of-sums logic circuit for Table 11.1

11.4　逻辑电路的化简

显然逻辑函数可以容易地表示为最小项之和或者最大项之积。但是，如此直接得到的逻辑电路往往不是最简的，因此，所使用的门电路数不是最少的。例如，根据逻辑表达式

$$F = \bar{A}\bar{B}D + \bar{A}BD + BCD + ABC \tag{11.25}$$

This type of expression is called a sum of products (SOP). Following this procedure, we can always find an SOP expression for a logic output given the truth table. A logic circuit that implements (11.21) directly is shown in Fig. 11.11.

A shorthand way to write an SOP is simply to list the row numbers of the truth table for which the output is logic 1. Thus, we can write

$$D = \sum m(0, 2, 6, 7) \tag{11.22}$$

in which m indicate that we are summing the minterms corresponding to the rows enumerated.

11.3.2   Product-of-sums implementation

Another way to write a logic expression for D is to concentrate on the rows of the truth table for which D is 0. For example, in Table 11.1, these are the rows numbered 1, 3, 4, and 5. Then, we write a logical sum that equals 0 for each of these rows. Each input variable or its inverse in included in each sum. In writing the sum for each row, we invert the logic variables that are 1 in that row. For example, the logical sum $(A + B + \overline{C})$ equals logic 0 only for row 1. Similarly, $(A + \overline{B} + \overline{C})$ equals logic 0 only for row 3. The sum terms that include all of the input variables (or their inverses) are called **maxterms**.

Finally, we write an expression for the output variable as the logical product of maxterms. For Table 11.1, it yields

$$D = (A + B + \overline{C})(A + \overline{B} + \overline{C})(\overline{A} + B + C)(\overline{A} + B + \overline{C}) \tag{11.23}$$

This type of expression is called a product of sum (POS). We can always find a POS expression for a logic output given the truth table. A circuit that implements equation (11.23) is shown in Fig. 11.12.

A shorthand way to write a POS is simply to list the row numbers of the truth table for which the output is logic 1. Thus, we write

$$D = \prod M(1, 3, 4, 5) \tag{11.24}$$

in which M indicates the maxterms corresponding to the rows enumerated.

11.4   Minimization of Logic Circuits

We have seen that logic functions can be readily expressed either as a logical sum of minterms or as a logical product of maxterms. However, direct implementation of either of these expressions may not yield the best circuit in terms of minimizing the number of gates required. For example, consider the logical expression

$$F = \overline{A}\,\overline{B}D + \overline{A}BD + BCD + ABC \tag{11.25}$$

直接实现逻辑电路，至少需要两个非门、四个与门以及一个或门。因此我们需要对该表达式进行化简。合并前两项，我们能够得到

$$F = \overline{A}D(\overline{B} + B) + BCD + ABC$$

由于 $\overline{B} + B = 1$，于是

$$F = \overline{A}D + BCD + ABC$$

当然，只有当 $B = 1$、$C = 1$ 且 $D = 1$ 时，才有 $BCD = 1$。因此，要么 $\overline{A}D = 1$，要么 $ABC = 1$，因为 $\overline{A} = 1$ 或 $A = 1$ 必有一个成立。因此，冗余的 BCD 项可以从式中移除，得到

$$F = \overline{A}D + ABC \tag{11.26}$$

这样只使用了一个非门、两个与门和一个或门就实现了这个表达式。

11.4.1 卡诺图

就如我们所看到的，逻辑表达式可以通过代数方法来进行化简。但是这种化简的过程效率很低。使用基于图形化简的卡诺图方法，我们能够简单地找到逻辑表达式最简的实现方式。

卡诺图是一种方格图。每个方格代表逻辑函数的一个最小项，也就是对应真值表中的一行。两变量、三变量和四变量的卡诺图示例如图 11.13 所示。两变量的卡诺图由四个方格组成，每个方格对应一个最小项。类似地，三变量的卡诺图有 8 个方格，四变量的卡诺图有 16 个方格。

在图 11.13 中每个最小项对应着一个格子。例如，在三个变量的卡诺图中，最小项 $\overline{A}B\overline{C}$ 对应的是上方右侧的方格，与真值表各行对应的位组合分别标在图的左边和上边。例如，在四变量卡诺图中，与真值表中 $ABCD$ 为 1101 的行相对应的是图中第三行第二个方格。第三行左边标为 11，第二列上边标为 01。因此，我们可以很容易地找到与任一项最小项或真值表中任一行相对应的方格。

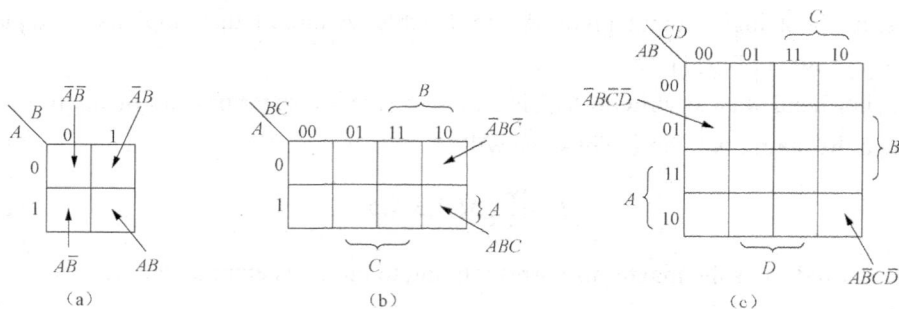

图 11.13　各方格对应最小项的卡诺图
（a）二变量；（b）三变量；（c）四变量

Fig. 11.13　Karnaugh maps showing the minterms corresponding to some of the squares

(a) Two variables; (b) Three variables; (c) Four variables

在四变量卡诺图的上方，位组合不是按照自然的二进制数的顺序排列，而是按照格雷码排列（即 00 01 11 10）。这样，任何两个相邻的方格之间只有一个变量不同，从而使相似的最小项集合在一起。例如，包含 A（而不是 $\overline{A}$）的最小项在每个图的下半部分。在四变量卡诺图中，包含 B 的最小项在图的中间两行，包含 AB 的最小项在图的第三行等。将相似最小项集合在一起是化简逻辑电路的关键。

Implemented directly, this expression would require two inverters, four AND gates, and one OR gate. Therefore, we need to simplify the above expression. Factoring the first pair of terms, we have

$$F = \overline{A}D + BCD + ABC$$

However, $\overline{B} + B = 1$, so we obtain

$$F = \overline{A}D(B + \overline{B}) + BCD + ABC$$

Of course, $BCD=1$ holds only if $B=1$, $C=1$ and $D=1$. In this case, either $\overline{A}D=1$ or $ABC=1$, because we must have either $\overline{A}=1$ or $A=1$. Thus, the term BCD is redundant and can be dropped from expression. Then, we get

$$F = \overline{A}D + ABC \tag{11.26}$$

Only one inverter, two AND gates, and one OR gate are required to implement this expression.

11.4.1 Karnaugh maps

As we have demonstrated, logic expressions can sometimes be simplified dramatically. However, the algebraic manipulations needed to simplify a given expression are often not readily apparent. By using a graphical approach known as the **Karnaugh map**, we will find it much easier to minimize the number of terms in a logic expression.

A Karnaugh map is an array of squares. Each square corresponds to one of the minterms of the logic variables or, equivalently, to one of the rows of the truth table. Karnaugh maps for two, three, and variables are shown in Fig. 11.13. The two-variable map consists of four squares, one corresponding to each of the minterms. Similarly, the three-variable map has eight squares, and the four-variable map has 16 squares.

The minterms corresponding to some of the squares are shown in Fig. 11.13. For example, for the three-variable map, the minterm $\overline{A}B\overline{C}$ corresponds to the upper right-hand square. Also, the bit combinations corresponding to the rows of the truth table are shown down the left-hand side and across the top of the map. For example, on the four-variable map, the row of the truth table for which the four-bit word $ABCD$ is 1101 corresponds to the square in the third row (i.e. the row labeled 11) and second column (i.e. the column labeled 01). Thus, we can readily find the square corresponding to any minterm or to any row of the truth table.

In case you are wondering about the order of the bit patterns along the side or top of the four-variable Karnaugh map (i.e. 00 01 11 10), notice that this is a two-bit Gray code. Thus, the patterns for squares with a common side differ in only one bit, so that similar minterms are grouped together. For example, the minterms containing A (ranther than $\overline{A}$) fall in the bottom half of each map. In the four-variable map, the minterms containing B are in the middle two rows, the minterms containing AB are in the third row, and so forth. This grouping of similar min terms is the key to simplifying logic circuits.

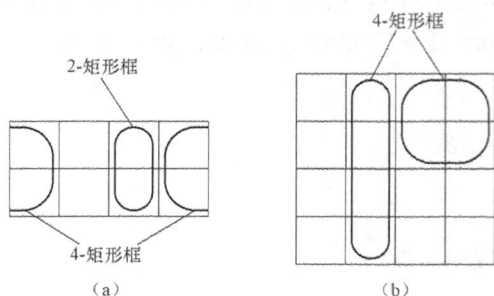

图 11.14 卡诺图中的矩形框

Fig. 11.14 Some cubes in Karnaugh map

包含一个共同边的两个方格被称为 2-矩形框。类似地，具有共同边的四个方格的矩形框被称为 4-矩形框。在选定矩形框时，图的上边和下边，左边和右边视为相邻。这样，在右边的方格与左边的方格相邻，上边的方格与下边的方格相邻。从而，在卡诺图的四个角上的四个方格形成一个 4-矩形框。图 11.14 给出了卡诺图中的几种矩形框。

为了得到逻辑函数的卡诺图，通常将使函数取 1 的方格标记为 1。将标有 1 的方格归入矩形框中。图 11.15 给出了一些两变量与项的图。

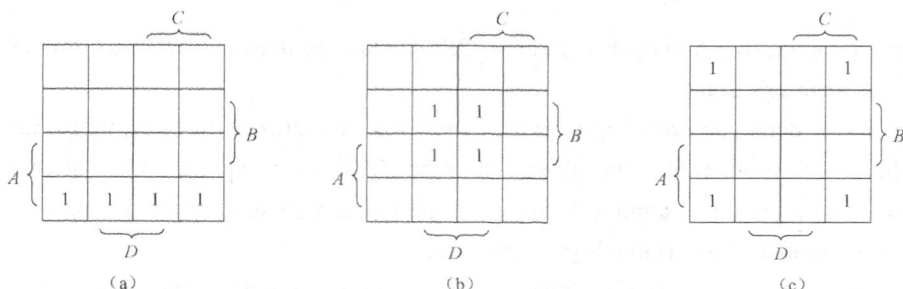

图 11.15 在四变量卡诺图中由 4-矩形框表示的两变量与项

（a） $A\overline{B}$ 的图；（b） BD 的图；（c） $\overline{B}\overline{D}$ 的图

Fig. 11.15 Products of two variables map into 4-cubes on a 4-variable Karnaugh map

(a) Map of $A\overline{B}$; (b) Map of BD; (c) Map of $\overline{B}\overline{D}$

由一个 16 个方格组成的四变量卡诺图中，单个逻辑变量或者其反变量的图为一 8-矩形框，一个两变量与项（例如 AB 或 $A\overline{B}$ ）的图为一个 4-矩形框，一个三变量与项的图为一个 2-矩形框。

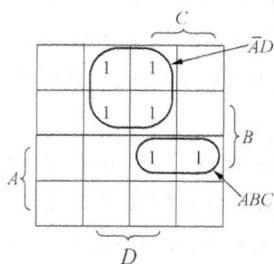

图 11.16 式（11.27）的卡诺图

Fig. 11.16 The Karnaugh map of (11.27)

图 11.16 给出了逻辑函数

$$F = \overline{A}\overline{B}\overline{C}D + \overline{A}\overline{B}CD + \overline{A}B\overline{C}D + \overline{A}BCD + ABC\overline{D} + ABCD \tag{11.27}$$

的卡诺图。图中的 4-矩形框对应于项 $\overline{A}D$ ，2-矩形框对应于项 ABC 。这些矩形框均是由标为 1 的方格组成的最大矩形框。于是可得到 F 的最简与或式表达式为

$$F = \overline{A}D + ABC \tag{11.28}$$

通过观察卡诺图，可以比较容易地得到含 1 的最大矩形框，所以可以很快地化简一个逻辑函数。

We call two squares that have a common edge a 2-cube. Similarly, four squares with common edges are called a 4-cube. In locating cubes, the maps should be considered to fold around from top to bottom and from left to right. Therefore, the squares on the right-hand side are considered to be adjacent to those on the left-hand side, and the top of the map is adjacent to the bottom. Consequently, the four squares in the map corners form a 4-cube. Some cubes are illustrated in Fig. 11.14.

To map a logic function, we place 1 in the squares for which the logic function takes a value of 1. Product terms that are labeled 1 are mapped into cubes. Fig. 11.15 gives some examples where product terms are mapped into cubes.

In a four-variable map consisting of 16 squares, a single logic variable or its inverse covers (maps into) an 8-cube. A product of two variables (such as AB or $A\overline{B}$) covers a 4-cube. A product of three variables maps into a 2-cube.

The Karnaugh map of the logic function

$$F = \overline{A}\,\overline{B}\,\overline{C}D + \overline{A}\,\overline{B}CD + \overline{A}B\overline{C}D + \overline{A}BCD + ABC\overline{D} + ABCD \tag{11.27}$$

is shown in Fig. 11.16. The squares containing 1 form a 4-cube corresponding to the product term $\overline{A}D$, plus a 2-cube corresponding to ABC. These are the largest cubes that cover the squares that are labeled 1 in the map. Thus, the minimum SOP expression for F is

$$F = \overline{A}D + ABC \tag{11.28}$$

Because it is relatively easy to spot the sets of largest cubes that cover the 1s in a Karnaugh map, we can quickly minimize a logic function.

参 考 文 献

［1］秦曾煌. 电工学：上册　电工技术. 北京：高等教育出版社，2011.

［2］秦曾煌. 电工学简明教程. 2 版. 北京：高等教育出版社，2007.

［3］唐介. 电工学（少学时）. 3 版. 北京：高等教育出版社，2009.

［4］邱关源. 电路. 5 版. 北京：高等教育出版社，2011.

［5］唐庆玉. 电工技术与电子技术：上册. 北京：清华大学出版社，2007.

［6］童诗白，华成英. 模拟电子技术基础. 4 版. 北京：高等教育出版社，2007.

［7］A R Hambley. Electrical Engineering Principles and Applications. 4th ed. 北京：机械工业出版社，2010.

［8］C K Alexander, M N O Sadiku. Fundamentals of Electric Circuits. 北京：清华大学出版社，2006.

［9］W H Hayt, Jr, J E Kemmerly, S M Durbin. Engineering Circuit Analysis. 6th ed. [S.l.] The McGraw-Hill Companies, 2002.

［10］J W Nilsson, S A Riedel. Electric Circuits. 8th ed. [S.l.] Prentice Hall, 2008.

［11］A S Sedra, K C Smith. Microelectronic Circuits. 6th ed. [S.l.] Oxford University Press, 2009.